DESIGN OF EARTHQUAKE-RESISTANT BUILDINGS

DESIGN OF EARTHQUAKE-RESISTANT BUILDINGS

Minoru Wakabayashi

McGRAW-HILL BOOK COMPANY

New York   St. Louis   San Francisco   Auckland
Bogotá   Hamburg   Johannesburg   London
Madrid   Mexico   Montreal   New Delhi
Panama   Paris   São Paulo   Singapore
Sydney   Tokyo   Toronto

Library of Congress Cataloging in Publication Data

Wakabayashi, Minoru, 1921–
  Design of earthquake-resistant buildings.

  Includes index.
  1. Earthquake resistant design.   I. Title.
TA658.44.W34   1986   693.8′52   84-29710

1234567890   DOC/DOC   898765

ISBN 0-07-067764-6

The editors for this book were Joan Zseleczky and Beatrice E. Eckes, the designer
was Naomi Auerbach, and the production supervisor was Sally Fliess. It was set in
Baskerville by The Saybrook Press, Inc.
Printed and bound by R. R. Donnelley & Sons Company.

CONTENTS

PREFACE

This book is intended for use by undergraduate and graduate students of structural engineering as well as by practicing structural engineers who are involved in the design and construction of building structures. Parts of the book will also be of interest to architects.

Structural problems which arise from earthquake action are not restricted to the geographical regions of intense earthquake activity, such as the area around the Pacific Ocean, but in fact occur in most parts of the world. This book is intended to provide information on recent world trends in earthquake engineering and should be applicable to earthquake-resistant design for any country or region. Although the book is not oriented to the situation in any particular country, regulations in the United States are often used as examples in the detailed discussions of earthquake-resistant design. These regulations are internationally well known and are often referred to by designers in other countries. Those who wish to study design specifications applying in a particular country should consult the literature of that country.

It seems that books dealing with earthquake-resistant structures are often specialized and therefore unsuitable for use as textbooks for the inexperienced reader. An attempt has been made in the present book to provide a simple, well-balanced, broad coverage of the information needed for the design of earthquake-resistant structures.

Chapter 1 first deals with the causes of earthquakes, seismic activity around the world, and damage caused by past major earthquakes. This information is of importance to architects as well as to structural engineers. Information on ground motion and its measurement is also contained in this chapter and is essential to the structural engineer for application in design practice.

Chapter 2 treats those elements of vibration theory which are closely related to earthquake-resistant design. Earthquake forces, as distinct from wind loads, act on a structure through vibration of the ground. The structural engineer should therefore be familiar with the vibration characteristics of a structure. This chapter has been written so that a reader with an undergraduate knowledge of mathematics and mechanics should easily be able to follow the development. Attention is drawn in Sec. 2.7 to the concept of aseismic safety, which is useful in evaluating structural behavior from the viewpoint of earthquake-resistant capacity.

Chapter 3 discusses the statical behavior of various structures under simulated earthquake loading. This forms the basis for earthquake-resistant design, which is dealt with in Chapter 4. In Chapter 3, strength, deformation capacity, and hysteretic behavior of members, connections, and systems are considered. Reinforced-concrete structures, steel structures, mixed structures, masonry structures, and wood structures are dealt with in turn.

Chapter 4 deals with methods of earthquake-resistant design. Detailed descriptions are first given of two methods: the static method of design, which is applied to most normal building structures; and the dynamic method of design, which is used for large-scale or important building structures. Section 4.5 takes up design questions which relate to earthquake-resistant capacity: how to select structural materials, structural forms, and framing systems. This section is particularly recommended to architects, since the earthquake-resistant capacity of a building structure is largely determined in the initial planning stages when decisions are taken with regard to structural form, layout, and materials. Section 4.7 deals with the design of equipment, facilities, and nonstructural elements such as cladding. These items have recently become of interest because of recorded damage during earthquake action.

Chapter 5 describes design methods for foundations.

Chapter 6 deals with the evaluation of the aseismic safety of existing building structures. Methods are also described for the repair of damaged structures and for the strengthening of building structures with inadequate earthquake resistance. These are quite new topics. Their importance has only recently been fully recognized, and much research work is still required.

It is my pleasure to acknowledge my gratitude to all who have helped me in the preparation of this text. My first thanks must go to Professors R. F. Warner, H. Tajimi, and Y. Kishimoto for their editorial guidance and encouragement. A number of present and former colleagues have helped in various ways. I owe deep gratitude for assistance and advice to Professors C. Matsui, T. Fujiwara, T. Nakamura, and S. Morino and to Mr. Y. Kishima and Dr. M. Nakashima. My gratitude is also extended to students who prepared figures and to Miss K. Rokuta, who typed the manuscript.

Minoru Wakabayashi

1

EARTHQUAKES AND GROUND MOTION

1.1 Earthquakes

1.1.1 Causes of Earthquakes

1.1.1.1 Plate Tectonics Of the various theories which have been proposed on the causes of earthquakes, the plate tectonics theory is now considered to be the most reliable. This theory tells us that the earth is covered by several layers of hard plates which act on each other to generate earthquakes. Hard tectonic plates, the *lithosphere*, sit on a comparatively soft *asthenosphere* and move as rigid bodies (Fig. 1-1). The plates measure about 70 km in thickness under the sea and twice that thickness under land. At the plate boundaries there are *midoceanic ridges, transform faults, island arcs*, and *orogenic zones*. At the midoceanic ridges, hot mantle flows up toward the surface of the earth and cools down, forming the plate, which expands horizontally. The tectonic plates pass each other at the transform faults and are absorbed back into the mantle at the orogenic zones. Earthquakes are often generated at subduction zones (Fig. 1-2) and in regions where the plates slip one against another.

Shown in Fig. 1-3 are the location of tectonic plates, the direction of

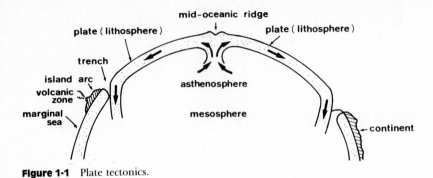

Figure 1-1   Plate tectonics.

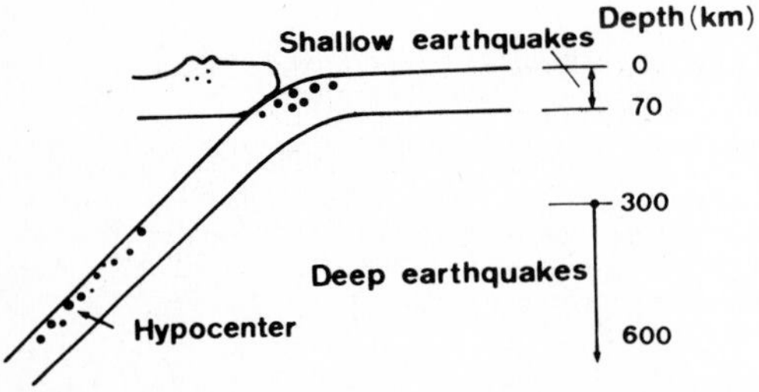

Figure 1-2   Idealized model of subduction zone and hypocenter.

plate movement, the distribution of ocean ridges, and so forth (Berlin, 1980; Bolt, 1978; Utsu, 1977). A comparison of this figure with a seismicity map (Fig. 1-11) adds credence to the plate tectonics theory.

An island arc is a chain of islands in the shape of an arc that is formed outside the marginal sea. Examples are the Kuril Islands, the Aleutian Islands, and the island chain of Japan. An island arc exhibits high seismicity potential and includes a volcano or volcanoes within its axis. Although the Pacific Ocean sides of Central and South America do not consist of islands, they are treated as island arcs since all their other characteristics are the same as those of island arcs. At the island arcs earthquakes are generated by the slip of one tectonic plate under the other. As illustrated in Fig. 1-2, these earthquakes are often deep (see Sec. 1.1.2.1).

1.1.1.2 Faults   Faults are formed when mutual slip of the rock beds occurs on a certain plane. Depending upon direction, the slippages are classified as follows:

1. *Dip slip.* Slippage takes place in a vertical direction.
   a. *Normal fault.* The upper rock bed slips downward (Fig. 1-4*a*).
   b. *Reverse fault.* The upper rock bed slips upward (Fig. 1-4*b*).

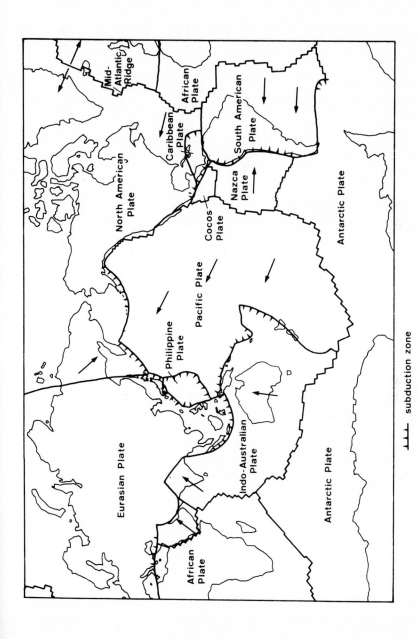

Figure 1-3 World map of tectonic plates. (*From Earthquakes—A Primer by B. A. Bolt. Copyright © 1978 by W. H. Freeman and Company, San Francisco. All rights reserved by B. A. Bolt.*)

⊥⊥⊥⊥ subduction zone

ᒧᒲᒧᒲ spreading zone

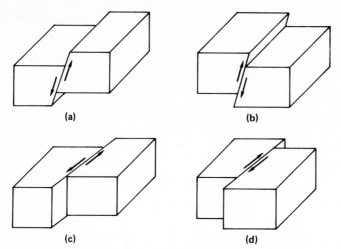

Figure 1-4   Main types of fault motion. *(a)* Normal fault. *(b)* Reverse fault. *(c)* Left lateral fault. *(d)* Right lateral fault.

2. *Strike slip.* Slippage takes place in a horizontal direction.
    a. Left lateral fault. As seen from one rock bed, the other rock bed slips toward the left (Fig. 1-4c).
    b. Right lateral fault. As seen from one rock bed, the other rock bed slips toward the right.

Actual faults are often a combination of the four types of slippages.

A fault that emerges at the surface of the earth because of an earthquake is called an *earthquake fault.* Earthquake faults are not formed by deep earthquakes.

The best-known example of an earthquake fault is the 300-km-long strike slip of 6.4 m at the San Andreas fault, which caused the San Francisco earthquake of 1906 (M = 8.3, where M is the magnitude on the Richter Scale shown in Sec. 1.1.3.2). During the Imperial Valley earthquake of 1940 (M = 7.1), a 60-km-long right lateral fault was created with a maximum slip of 5 m (see Fig. 1-12 below).

One of the most famous faults in Japan was created by the Nobi earthquake (M = 8.4) in 1891. It is 80 km long and showed a 6-m vertical slip and a 2- to 4-m horizontal slip. The Kansu earthquake of 1920 (M = 8.5) in China created a left lateral fault 200 km long.

Generally speaking, the length and width of a fault are comparable when the fault is created by relatively small earthquakes of M < 6 (see Sec. 1.1.3.2), but such earthquakes rarely form earthquake faults. Faults are longer when earthquakes are greater (Sec. 1.1.3.2).

Faults are causes rather than results of earthquakes. An earthquake is caused by a fault in the following ways:

1. Strain that has accumulated in the fault for a long period of time reaches its limit (Fig. 1-5a).

2. Slip occurs at the fault and causes a rebound (Fig. 1-5b).

3. A push-and-pull force acts at the fault (Fig. 1-5c).

4. This situation is equivalent to two pairs of coupled forces suddenly acting (Fig. 1-5d).

5. This action causes radial wave propagation.

The moment of each couple is called the *earthquake moment* or *seismic moment* (Kasahara, 1981). The seismic moment is defined as the rigidity of the rock times the area of faulting times the amount of slip. Recently it has been used as a measure of earthquake size (see Sec. 1.1.3.2).

Active faults are faults that have undergone deformation for the past several hundred thousand years and will continue to do so in the future. They have been found by geological and topographical surveys and aerial photographs. Since earthquakes often occur at active faults, when designing an important structure such as a nuclear power plant to resist seismic forces, the distance from a nearby active fault or faults to the building site, seismic activity, and other factors related to the fault are taken into account in predicting seismic motion of the ground.

The famous San Andreas fault in California reveals itself on land between Point Arena and the Gulf of California. A right lateral fault, it occurs where the Pacific plate slips to the north against the North American plate. It caused the Fort Tejon earthquake (1857), the San Francisco earthquake (1906), and other earthquakes.

Average slip velocity at an active fault varies. The highest velocities,

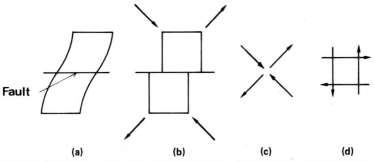

Figure 1-5   Earthquake mechanism. *(a)* Before slip. *(b)* Rebound due to slip. *(c)* Push-and-pull force. *(d)* Double couple.

those of the San Andreas fault and the Nankai trough of Japan, are 10 to 100 mm/year. A slippage of 3 m during one earthquake therefore means that earthquakes occur at intervals of 30 to 100 years at these faults. Some active faults, such as the San Andreas fault, are always moving; others, such as some faults in Japan, move only when an earthquake occurs.

1.1.2 Earthquakes and Seismic Waves

1.1.2.1 Epicenter   The point where the seismic motion originates is called the *focus, center,* or *hypocenter* of the earthquake; and the projection of the focus onto the surface of the earth is the *epifocus* or *epicenter*. The distances from the focus and the epicenter to the point of observed ground motion are called the *focal distance* and the *epicentral distance,* respectively.

Seismic destruction propagates from the focus through a limited region of the surrounding earth body, which is called the *focal region*. The larger the earthquake, the larger the focal region.

Earthquakes are classified as shallow, intermediate, and deep, depending on the depths of their foci. Limiting depths are often set at 70 km and 300 km.

1.1.2.2 Seismic Waves   Two types of seismic wave travel from the foci in the earth body: the *body wave* and the *surface wave*. The body wave, which propagates in an infinite continuum, is a twofold P wave and S wave. The P wave is often called the longitudinal wave or the compressive wave; it propagates in the same direction as its own vibration. The S wave is called the transverse wave or the shear wave; it propagates in a direction perpendicular to its vibration.

The propagation velocities of the P wave, V_p, and the S wave, V_s, are expressed as follows:

$$V_p = \left[\frac{E}{\rho} \frac{1 - \nu}{(1 + \nu)(1 - 2\nu)} \right]^{\frac{1}{2}} \tag{1-1}$$

$$V_s = \left(\frac{G}{\rho} \right)^{\frac{1}{2}} = \left[\frac{E}{\rho} \frac{1}{2(1 + \nu)} \right]^{\frac{1}{2}} \tag{1-2}$$

where E = Young's modulus
$\quad\quad G$ = shear modulus
$\quad\quad \rho$ = mass density
$\quad\quad \nu$ = Poisson ratio

For any material $V_p > V_s$, and if the Poisson ratio for the earth body is taken to be 0.25, then $V_p = \sqrt{3}\ V_s$ is obtained from Eqs. (1-1) and (1-2). Near the surface of the earth, $V_p = 5$ to 7 km/s and $V_s = 3$ to 4 km/s.

Surface waves propagate on the earth's surface and are detected more often in shallow earthquakes. They are mainly classified into two kinds: L waves (Love waves) and R waves (Rayleigh waves). An L wave takes place in stratified formations and vibrates in a plane parallel to the earth's surface and perpendicular to the direction of wave propagation. An R wave vibrates in a plane perpendicular to the earth's surface and exhibits an elliptic movement. Its velocity is smaller than but nearly equal to that of an S wave.

A P wave arrives at an observation station earlier than an S wave because its velocity is higher. In the earthquake accelerograms of Fig. 1-6 the P wave is recorded for some time before the S wave arrives.

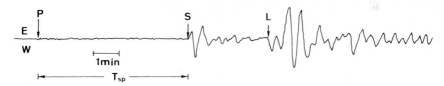

Figure 1-6 Earthquake accelerogram. The arrivals of the P, S, and surface (L) waves are marked. Time increases from left to right. *(Courtesy of T. M. Mikumo.)*

The time interval between the arrival at the observation station of a P wave and an S wave is called the *duration of preliminary tremors*, T_{sp}. If the two waves travel along the same route and have a constant velocity, the following equation gives the duration of the preliminary tremors:

$$T_{sp} = \left(\frac{1}{V_s} - \frac{1}{V_p} \right) \Delta \tag{1-3}$$

in which Δ is the distance from the focus to the observation point. This means that the epicenter can be located and the depth of the focus can easily be obtained graphically if earthquake records are made at least at three different observation points.

1.1.3 Scale and Intensity of Earthquakes

1.1.3.1 Intensity Scale   An *intensity scale* is the scale of ground-motion intensity as determined by human feelings and by the effects of ground motion on structures and on living things. It is graded according to intensity.

Proposed intensity scales included the Gastaldi Scale (1564) and the Pignafaro Scale (1783). The Rossi-Forel Scale (1883), which has 10 grades, is still used in some parts of Europe. The Mercalli-Cancani-Sieberg Scale, developed from the Mercalli (1902) and Cancani Scales (1904), is still widely used in western Europe. In 1931 F. Neumann modified the Mercalli-Cancani-Sieberg Scale, proposing a 12-grade Modified Mercalli (MM) Scale, which has now been widely adopted in North America and other parts of the world (see Table 1-1). Other intensity scales are the 12-grade Medvedev-Sponheuer-Karnik (MSK) Scale (1964), which is intended to unify intensity scales internationally, and the 8-grade scale of the Japanese Meteorological Agency (JMA).

Intensity scales are established on the basis of visible phenomena and human feelings as indicated in Table 1-1. Therefore they bear no specific relation to the maximum acceleration of ground motion, and correlation among different intensity scales is not necessarily clear. Figure 1-7 is one attempt to correlate intensity scales (AIJ, 1981).

If seismic intensities at various points for a small earthquake are plotted on a map, the ideal isoseismal pattern shows a bell shape. The shape will be as in Fig. 1-8 if the causative fault is several hundred kilometers long (Housner, 1969). In reality, however, the isoseismal pattern is dependent upon conditions at the epicenter, the route of the seismic wave from the focus to the observation station, geological conditions at the observation points, and other influences, and its shape is more complex.

1.1.3.2 Magnitude   The size of an earthquake is closely related to the amount of energy released. The magnitude M defined by Richter in 1935

TABLE 1-1 Abridged Modified Mercalli Earthquake Intensity Scale

Intensity value	Description
I	Not felt except under exceptionally favorable circumstances
II	Felt by persons at rest
III	Felt indoors; may not be recognized as an earthquake
IV	Windows, dishes, and doors disturbed; standing motor cars rock noticeably
V	Felt outdoors; sleepers wakened; doors swing
VI	Felt by all; walking unsteady; windows and dishes broken
VII	Difficult to stand; noticed by drivers; fall of plaster
VIII	Steering of motor cars affected; damage to ordinary masonry
IX	General panic; weak masonry destroyed, ordinary masonry heavily damaged
X	Most masonry and frame structures destroyed with foundations; rails bent slightly
XI	Rails bent greatly; underground pipes broken
XII	Damage total; objects thrown into the air

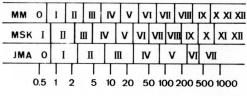

Figure 1-7 Relationship between different types of intensity scales and maximum acceleration. [*From AIJ*, Data for Earthquake Resistant Design of Buildings, *Tokyo, 1981 (in Japanese).*]

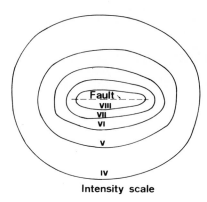

Intensity scale

Figure 1-8 Contour lines of equal intensity of ground shaking.

is often used to express earthquake size. The value of M is given by Eq. (1-4) when the Wood-Anderson type of seismometer shows a maximum amplitude of A μm at a point 100 km from the epicenter:

$$M = \log A \tag{1-4}$$

However, a standard seismometer is not always set at a point 100 km from the epicenter, in which case one may use Eq. (1-5):

$$M = \log A - \log A_0 \tag{1-5}$$

where A is the maximum recorded trace amplitude for a given earthquake at a given distance and A_0 is that for a particular earthquake selected as a standard.

Since the damping of a seismic wave is dependent on the underground structure, the formula for the magnitude derived for California cannot be directly applied to determine the value for other areas and must be modified. For instance, for Californian earthquakes, Eq. (1-6),

$$M = \log a + 3 \log \Delta - 2.92 \tag{1-6}$$

and for Japanese earthquakes, Eq. (1-7),

$$M = \log a + 1.73 \log \Delta - 0.83 \tag{1-7}$$

have been derived. Here a, μm, is the ground amplitude, and Δ, km, is the epicentral distance. These equations are more generally applicable than Eq. (1-5) because the maximum trace amplitude is replaced by the ground amplitude a, and they may be used for any type of seismograph (Kasahara, 1981).

Also, for earthquakes at teleseismic distances, surface waves often predominate over body waves, which have different damping characteristics, and different formulas are needed to determine the value of M from the amplitudes of the surface waves. Since the magnitude of an earthquake depends on the paths of the seismic wave, the underground structure near the observation station, and various other conditions, the magnitude determined from the data of one observation station differs from that of another station. The difference, in general, is as much as one-fourth greater than the average value.

Richter's magnitude is called a *local magnitude;* it cannot be applied to an earthquake with a long epicentral distance. For earthquakes at teleseismic distances, a surface-wave magnitude with an approximately 20-s period is sometimes used instead.

The seismic intensity I is large for a large magnitude M and a short epicentral distance r. Many relationships have been developed for these three quantities, among them that of Esteva and Rosenblueth (1964):

$$I = 8.16 + 1.45M - 2.46 \ln r \tag{1-8}$$

in which I is measured on the MM Scale and r is in kilometers. This equation cannot be used for cases in which r reaches the same order as the focal region (Utsu, 1977). Figure 1-9 illustrates the above relation.

Earthquakes of larger magnitude occur less frequently than those of smaller magnitude. Equation (1-9) is an empirical relation between magnitude M and the number N of earthquakes of magnitude larger than M

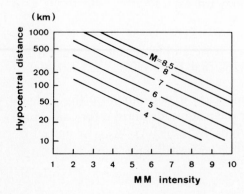

Figure 1-9 Relationship between hypocentral distance, magnitude, and seismic intensity.

which occur at a particular location over a unit period of time (Gutenberg and Richter, 1956):

$$\log N = A - (b \cdot M) \tag{1-9}$$

The constants A and b depend solely on the location under consideration. Dowrick (1977) gives values of A and b for various locations in the world.

Figure 1-10 shows the distribution of the magnitude of earthquakes in the neighborhood of Japan, including inland areas and offshore areas within 200 km of the shoreline, which had epicenters shallower than 60 km. The figure indicates that Eq. (1-9) holds fairly well in the region (Utsu, 1977). Housner (1969; 1970) gives the relation between the magnitudes of earthquakes expected in California and the number of earthquakes which had occurred in the past 100 years.

The length of the earthquake fault L in kilometers is related to the magnitude (Tocher, 1958):

$$M = (0.98 \cdot \log L) + 5.65 \tag{1-10}$$

The slip in the fault U in meters is also related to the magnitude (Chinnery, 1969):

$$M = (1.32 \cdot \log U) + 4.27 \tag{1-11}$$

A similar equation is given by Iida (1965).

1.1.3.3 Energy   Part of the strain energy released by an earthquake is dispersed from the focal region as seismic-wave motion. The rest becomes potential energy, which enables crustal deformation to take place and energy to be absorbed in the destruction of rocks and slippage at faults.

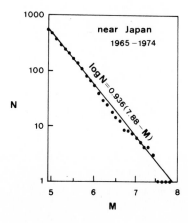

Figure 1-10 Relationship between magnitude and number of occurrences in Japan, 1965–1974. (*From T. Utsu*, Seismology, *Kyoritsu Shuppan Co., Tokyo, 1977.*)

Gutenberg and Richter (1956) give the following relationship between the energy of the seismic wave E and the magnitude M:

$$\log E = 4.8 + 1.5M \tag{1-12}$$

Several similar equations have been proposed. In Eq. (1-12), the value E is given in joules. From the equation, it is noted that the energy increases by about 32 times for an increase of 1 in the magnitude and 1000 times for an increase of 2 in M.

Recently, earthquake moment has also been used as a measure of earthquake size.

1.1.4 Seismic Activity

1.1.4.1 World Seismic Activity   A map with plots of earthquake epicenters is called a *seismicity map;* it usually shows all the earthquakes with a magnitude larger than a specified value which have occurred in a certain period of time. Figure 1-11 is an example which shows the epicenters of earthquakes of a magnitude of 7 or larger during the period from 1900 to 1980. It suggests that the following areas are often subjected to earthquakes:

1. Circum-Pacific seismic zone, including the Pacific side of South America, Central America, and North America, the Aleutian Islands, the Kamchatka Peninsula, Japan, Indonesia, and New Zealand

2. Eurasian seismic zone, stretching from southeast Asia through the Middle East to the Mediterranean Sea

3. Midoceanic ridge and the area where the ridge reveals itself on land

4. Parts of China, North America, the Middle East, and other continents

Most of the areas included in item 1 are either island arcs where tectonic plates submerge, island-arc-like areas such as the coastal area of Central and South America, or transform-fault areas such as the coastal area of North America. A comparison of Fig. 1-3 with Fig. 1-11 helps justify the plate tectonics theory, which explains that earthquakes are produced where tectonic plates act on each other.

As shown in Fig. 1-11, areas where earthquakes occur frequently are often distributed in long, narrow strips. These areas are therefore called *seismic zones*. Recent large earthquakes of the class $M = 8$ have occurred in the trenches of island arcs, filling up areas where no (or very few) earthquakes had taken place in recent years. The term *seismicity gap* is

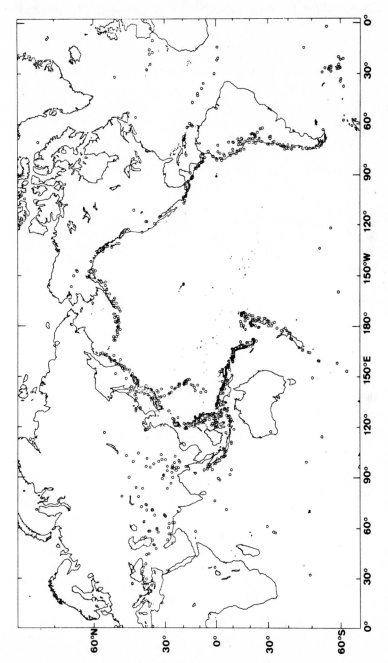

Figure 1-11 Seismicity map of the world showing epicenters for all earthquakes with magnitudes of 7 or larger that occurred between 1900 and 1980. (*Drawn from computer data of the Information Processing Center for Disaster Prevention Studies, Disaster Prevention Research Institute, Kyoto University, Kyoto, Japan.*)

applied to such regions in which no recent earthquakes have taken place and where earthquakes will possibly take place in the comparatively near future.

Table 1-2 presents details of recent large earthquakes in various parts of the world. It will be clear that damage and loss of life depend on population density, form of building construction, etc., and not solely on earthquake intensity. It is also difficult to compare directly historical epicenter maps of an old area such as China or Japan with those of a new area such as North America.

1.1.4.2 North America   Along the shoreline of California there runs the San Andreas fault, where many earthquakes have occurred (Fig. 1-12). This fault is a transform fault of the North American plate and the Pacific plate, with the latter moving 50 to 60 mm/year relative to the former in a north by northwest direction. The motion is essentially right lateral, but regional features are complicated (Kasahara, 1981). The San Francisco earthquake (1906), the Imperial Valley earthquake (1940), and the Parkfield earthquake (1966) occurred along the San Andreas fault, while the Kern County earthquake (1952) and the San Fernando earthquake (1971) were associated with branches of the fault (see Table 1-2). The fault has produced about 30 major earthquakes. According to the seismic-probability map shown in Fig. 1-13, earthquakes have also occurred in the northern part of New York State and in Missouri as well as in California and Nevada; one earthquake in Missouri (1811−1812) caused extraordinary geographical phenomena. This earthquake had areas of structural damage 5 times larger than that of the 1906 San Francisco earthquake (Nuttli, 1982). Such earthquakes, which are referred to as *interplate earthquakes*, are not associated with a plate boundary, and their cause is not explained by the plate model (Berlin, 1980).

1.1.4.3 Central and South America   Because the Cocos plate and the Nazca plate are submerging on the Pacific Ocean side of Central and South America, respectively, this island-arc-like area has high seismicity. In recent years, destructive earthquakes have occurred: the Mexico City earthquakes (1957 and 1973), the Guatemala earthquake (1976; 23,000 people killed), and the Peru earthquake (1970; 70,000 killed). (See Table 1-2.) In Peru and Chile, extremely large earthquakes of the class $M = 8$ have initiated tsunami, which in turn have caused damage in areas as remote as Hawaii and Japan.

1.1.4.4 Japan   The Pacific plate and the Philippine plate submerge along the Pacific Ocean side of the Japanese islands. Seismicity is therefore very high in the area between the islands and the ocean ridge.

TABLE 1-2 Recent World Large Earthquakes

Area	Country	Year	Magnitude (Richter)	Region	Deaths	Comments
North America	Mexico	1957	7.9	Mexico City	68	
	United States	1811–1812	8	New Madrid, Mo.	1	
		1906	8.3	San Francisco, Calif.	700	
		1940	7.1	Imperial Valley, Calif.	8	
		1949	7.1	Olympia, Wash.	8	
		1952	7.7	Kern County, Calif.	12	
		1959	7.1	Hebgen Lake, Mont.	28	Tsunami
		1964	8.4	Prince William Sound, Alaska	131	Tsunami
		1971	6.5	San Fernando, Calif.	65	
Central America	Guatemala	1976	7.5		23,000	
	Nicaragua	1972	6.2	Managua	5,000	
South America	Chile	1960	8.3	Off coast of central Chile	1,743	Hawaii, 61, and Japan, 119, by tsunami
	Peru	1970	7.6	Near coast of northern Peru	70,000	
	Venezuela	1965	6.5	Caracas	266	
Asia	China	1920	8.5	Kansu	100,000	
		1925	7.1	Yunnan	6,500	
		1976	7.6	Tangshan	650,000	
	India	1950	8.6	India-China border	574	
	Japan	1891	8.4	Nobi	7,273	
		1896	7.6	Sanriku	27,122	By tsunami
		1923	7.9	Kanto	143,000	Mainly by fire
		1927	7.5	Kitatango	2,925	
		1933	8.3	Sanriku	3,008	By tsunami
		1943	7.4	Tottori	1,083	
		1944	8.0	Tonankai	998	
		1945	7.1	Mikawa	1,961	
		1946	8.1	Nankaido	1,432	
		1948	7.3	Fukui	3,895	
		1964	7.5	Niigata	26	
		1968	7.9	Tokachi-Oki	49	
		1978	7.4	Miyagiken-Oki	27	
South Pacific area	New Zealand	1931	7.9	Hawke's Bay		
Middle East	Iran	1929	7.1	Iran-U.S.S.R. border	3,253	
		1962	7.3	Northwest Iran	12,000	
		1968	7.4		11,000	
	Turkey	1939	8.0	Erzincan	23,000	
		1944	7.4		4,000	
		1976	7.3	Northwest Iran–U.S.S.R. border	5,000	
Europe	Italy	1908	7.5	Messina	120,000	
		1915	7.5	Avezzano	35,000	
		1976	6.5	Fruili	968	
	Romania	1977	7.2	Vrancea	2,000	
	Yugoslavia	1963	6.0	Skopje	1,200	
		1979	7.3	Montenegro	121	
Africa	Morocco	1960	5.9	Agadir	14,000	

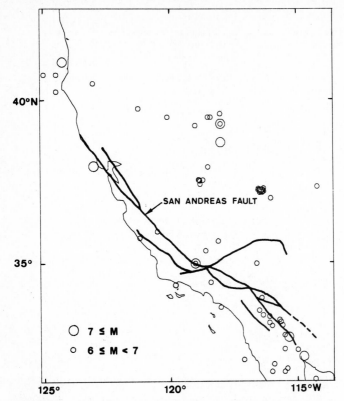

Figure 1-12 Seismicity map of California showing epicenters for all earthquakes with magnitudes of 6 or larger that occurred between 1900 and 1980. (*Drawn from computer data of the Information Processing Center for Disaster Prevention Studies. Diaster Prevention Research Institute, Kyoto University, Kyoto, Japan.*)

As indicated in Fig. 1-14*a*, large earthquakes of a magnitude greater than 8 tend to be located on the east side (sea area) of the islands. Since these earthquakes are caused by the submerging of the tectonic plates, they are often deep, as illustrated by Fig. 1-14*b*.

Smaller earthquakes with their foci at a depth of 20 km or less in the inland area of Japan are also frequent. They are destructive because they occur directly beneath cities. In the Kanto earthquake of 1923 ($M = 7.9$), 143,000 lives were lost, mostly because of fires triggered by the earthquake. From 1000 to 4000 lives were lost in the Tottori earthquake (1943; $M = 7.4$), the Tonankai earthquake (1944; $M = 8.0$), the Nankaido earthquake (1946; $M = 8.1$), and the Fukui earthquake (1948; $M = 7.3$). (See Table 1-2.)

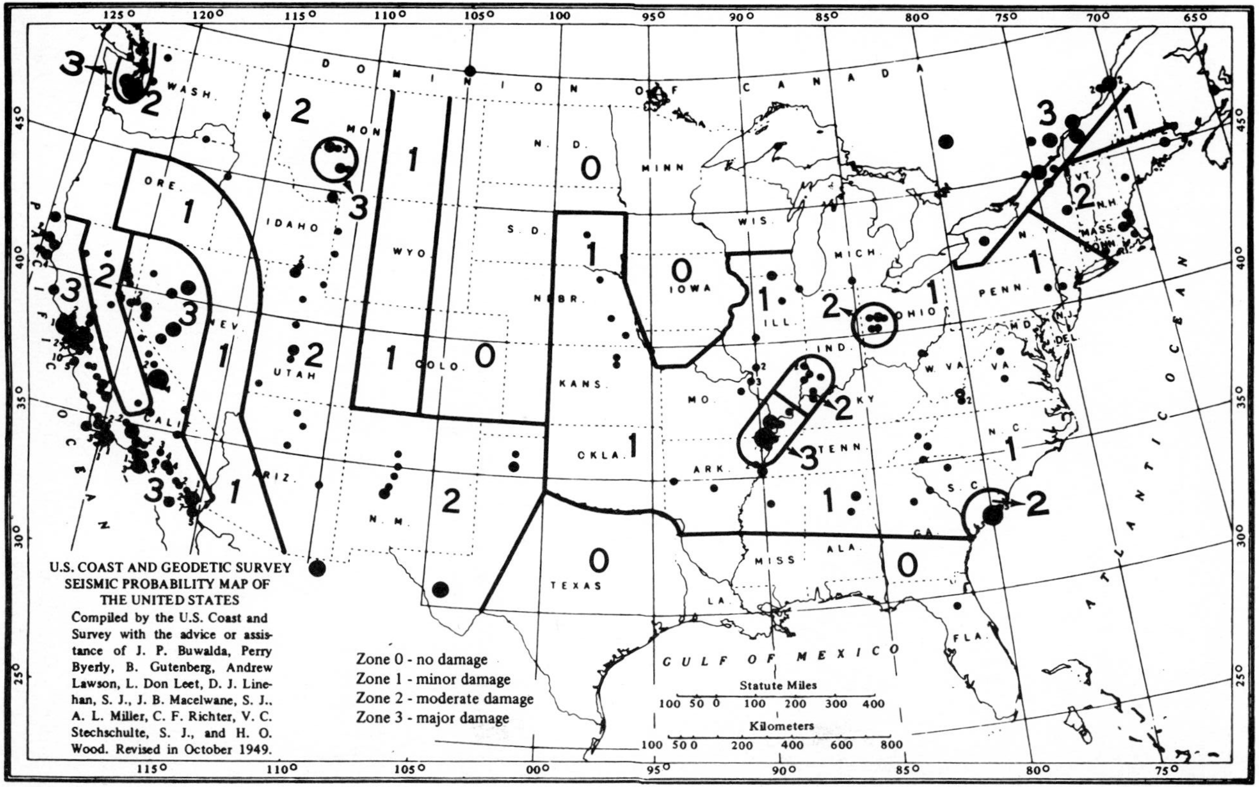
U.S. COAST AND GEODETIC SURVEY
SEISMIC PROBABILITY MAP OF
THE UNITED STATES
Compiled by the U.S. Coast and Survey with the advice or assistance of J. P. Buwalda, Perry Byerly, B. Gutenberg, Andrew Lawson, L. Don Leet, D. J. Linehan, S. J., J. B. Macelwane, S. J., A. L. Miller, C. F. Richter, V. C. Stechschulte, S. J., and H. O. Wood. Revised in October 1949.
Zone 0 - no damage
Zone 1 - minor damage
Zone 2 - moderate damage
Zone 3 - major damage
Statute Miles
100 50 0 100 200 300 400
Kilometers
100 50 0 200 400 600 800
D O M I N I O N O F C A N A D A
P A C I F I C O C E A N
A T L A N T I C O C E A N
G U L F O F M E X I C O
WASH.
ORE.
CALIF.
NEV.
IDAHO
MON.
WYO.
UTAH
ARIZ.
COLO.
N. M.
N. D.
S. D.
NEBR.
KANS.
OKLA.
TEXAS
MINN.
IOWA
MO.
ARK.
LA.
WIS.
ILL.
MICH.
IND.
OHIO
KY.
TENN.
MISS.
ALA.
GA.
FLA.
S. C.
N. C.
VA.
W. VA.
MD.
DEL.
N.J.
PENN.
N. Y.
VT.
N.H.
MASS.
CONN.
125°
120°
115°
110°
105°
100
95°
90°
85°
80°
75°
70°
65°
45°
40°
35°
30°
25°

Figure 1-13 Seismic-probability map of the United States. [*From G. W. Housner, Strong ground motion, in R. L. Wiegel (ed.),* Earthquake Engineering, *Prentice-Hall, Inc., Englewood Cliffs, N.J., 1970, pp. 75–91.*]

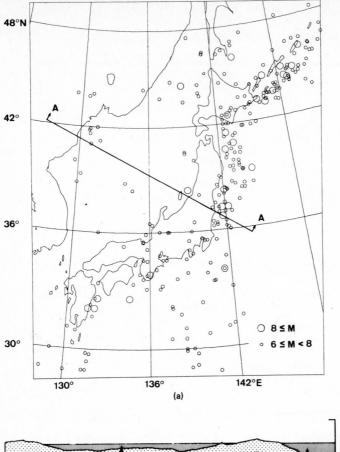

(a)

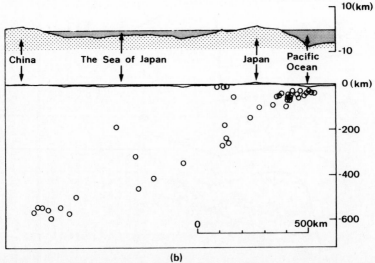

(b)

Figure 1-14 Large earthquakes in Japan with magnitudes of 6 or more that occurred between 1900 and 1980. (*Drawn from computer data of the Information Processing Center for Disaster Prevention Studies. Disaster Prevention Resarch Institute, Kyoto University, Kyoto, Japan.*)

1.1.4.5 South Pacific Area   In New Zealand seismicity in the eastern part of North Island and the southern part of South Island is as high as in southern California (Gutenberg and Richter, 1956). (See Fig. 1-15.) The largest earthquake experienced in New Zealand took place at Hawke's Bay in North Island in 1931 ($M = 7.9$) with damage to such cities as Napier and Hastings.

The Philippines suffered a large earthquake in 1976 with 4000 people killed.

1.1.4.6 Asia   Continental China is part of the Eurasian plate. The Pacific and Philippine plates push against it from the east, while the Indo-Australian plate pushes from the south, producing a very complicated stress field. Active faults of large size have been created, resulting in a number of in-plate earthquakes (Fig. 1-16).

Historical records in China tell us of large, destructive earthquakes: Shensi earthquake (1556; $M = 8$; 830,000 killed), Shantung earthquake (1668; $M = 8.5$), Kansu earthquake (1920; $M = 8.5$; 100,000 killed). As a

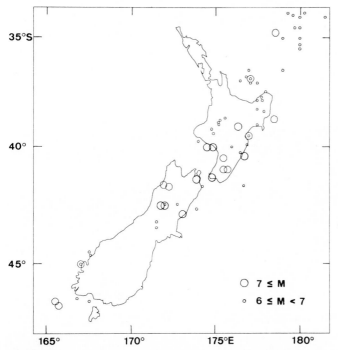

Figure 1-15   Large earthquakes in New Zealand: all earthquakes with magnitudes of 6 or larger that occurred between 1900 and 1980 and large historical shallow earthquakes. (*Drawn from computer data of the Information Processing Center for Disaster Prevention Studies. Disaster Prevention Resarch Institute, Kyoto University, Kyoto, Japan.*)

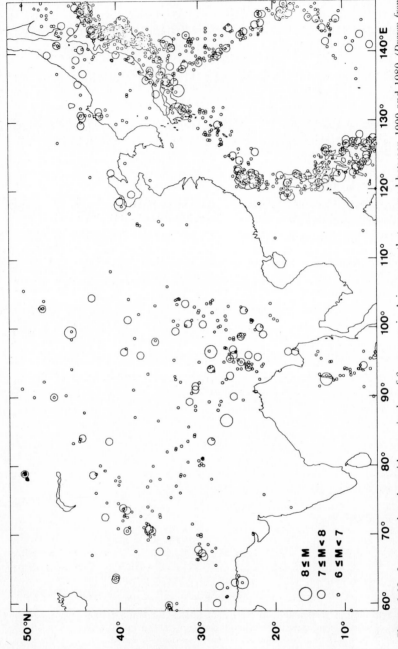

Figure 1-16 Large earthquakes with magnitudes of 6 or more in Asian areas that occurred between 1900 and 1980. (*Drawn from computer data of the Information Processing Center for Disaster Prevention Studies. Disaster Prevention Resarch Institute, Kyoto University, Kyoto, Japan.*)

result of the more recent Tangshan earthquake (1976; $M = 7.6$) the cities of Tangshan and Fengnan were almost completely destroyed, and as many as 650,000 lives were lost (Table 1-2).

Taiwan has seismological conditions similar to those in Japan and has high seismicity (Ming-Tung, 1965).

In the region around the Himalayan Mountains, large earthquakes of the class $M = 8$ have taken place with considerable loss of life. Eastern India, especially Assam and its neighborhood, has also been hit by many earthquakes. The Assam earthquake of 1897 damaged a wide area, while the India-China border earthquake of 1950 ($M = 8.6$) killed many people.

There have been many large earthquakes in the Soviet Union, particularly in the Kirghiz Republic and the Tadzik Republic.

1.1.4.7 Middle East   Some parts of Iran and Turkey have also experienced large earthquakes. Many people were killed by an earthquake in northwestern Iran (1962) and in the Erzincan earthquake (1939) and the Muradiye earthquake (1976), both in Turkey.

1.1.4.8 Europe   Countries in Europe with relatively high seismicity include Italy, Greece, Yugoslavia, and Romania (Fig. 1-17). In Italy, relatively small but shallow earthquakes, such as the Messina earthquake (1908; $M = 7.5$; 120,000 killed) and the Avezzano earthquake (1915; $M = 7$; 35,000 killed), often cause large damage. The Skopje earthquake (1963; $M = 6.0$) and the Montenegro earthquake (1979; $M = 7.3$) in Yugoslavia and the Vrancea earthquake (1977; $M = 7.2$) in Romania caused great damage as they occurred near cities. These were all shallow earthquakes, but they were not very large (Table 1-2).

1.1.4.9 Africa   There have been relatively very few earthquakes in the African continent, with some exceptions in Morocco, Algeria, and Tunisia.

1.2 Measurement of Earthquakes

1.2.1 Seismometer

1.2.1.1 Principle of the Seismograph   The principle of the seismograph is that ground motion is measured by the vibration record of a simple pendulum hanging from a steady point (Fig. 1-18; Kanai, 1969a).

In the system shown in Fig. 1-18, the displacement v of the pendulum is proportional to ground motion v_g, $v \propto v_g$, if the natural period of the pendulum is long relative to the period of ground motion and if an appropriate damping coefficient for the pendulum is chosen. The re-

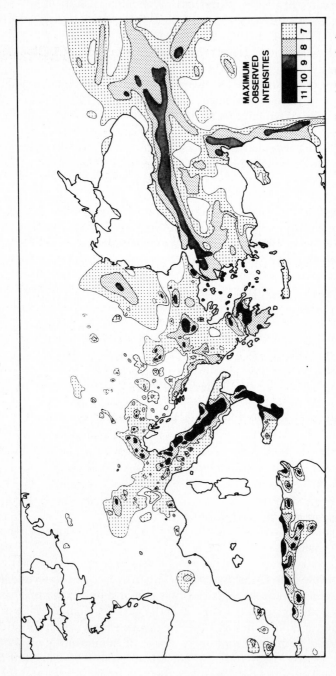

Figure 1-17 Maximum-observed-intensity map of Europe. *(From the map compiled by V. Karnik, C. Radu, G. Polonik, and D. Prochazkova. Courtesy of A. Cismigiu.)*

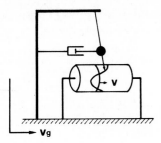

Figure 1-18 Seismograph.

corded displacement can thus be expressed in terms of ground motion times a constant. This type of seismograph is called either a *displacement seismograph* or a *long-period seismograph*. If the period of the pendulum is set short enough relative to that of ground motion, by means of an appropriate value of the pendulum's damping coefficient, $v \propto \ddot{v}_g$ is obtained. This means that ground acceleration can also be recorded by the seismograph. This type of seismograph is called an *acceleration seismograph* or a *short-period seismograph*. If the natural period of the pendulum is set close to that of ground motion and if the value of the damping coefficient of the pendulum is large enough, then $v \propto \dot{v}_g$. Ground velocity can be then determined. This type is called a *velocity seismograph*.

For a long-period seismograph, the pendulum must be very long in the system shown in Fig. 1-18 in order to achieve a long period. Instead of using a long pendulum, a horizontal pendulum (Fig. 1-19*a*) or an inverted pendulum (Fig. 1-19*b*) can be used to satisfy the condition.

1.2.1.2 Structure of the Seismograph   The movement of the pendulum can be amplified by mechanical, optical, or electromagnetic means. An amplification of several hundredfold can be achieved mechanically. With optical instrumentation amplification can be increased several thousandfold, and with electromagnetic techniques as much as several millionfold (Kanai, 1969*a*).

As stated earlier, the damping coefficient has to be appropriately chosen to achieve proportionality between the displacement of the pen-

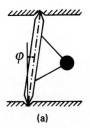

Figure 1-19 Pendulums for long-period seismographs. *(a)* Horizontal pendulum. *(b)* Inverted pendulum.

(a)   **(b)**

dulum and the displacement, velocity, or acceleration of ground motion. To this end, dampers can be operated by air, oil, or an electromagnet.

1.2.2 Strong-Motion Accelerograph

For purposes of seismic engineering, strong earthquakes must be recorded. Strong-motion accelerographs are made for this use (Cherry, 1974; Halverson, 1965; Hudson, 1970).

The recorder of a strong-motion accelerograph is normally at rest until ground acceleration exceeds a preset value and thereby triggers the measurement of any strong earthquake. The earthquake record can be made in three components of the vibration, two horizontal and one vertical.

The many types of strong-motion accelerographs include the U.S. Coast and Geodetic Survey Standard, Akashi SMAC B/B2, Teledyne AR-240, Teledyne RFT-250, Teledyne RMT-280, New Zealand M02, U.S.S.R., U.A.R., and so forth. In general, these accelerographs have the following characteristics:

1. The period and damping of the pickup is 0.06 to 25 cycles/s, selected so that the response is proportional to the acceleration of the ground.

2. The preset starting acceleration is approximately 0.005 g, and the accelerographs are sensitive to an acceleration from 0.001 to 1.0 g.

3. The average starting time is 0.05 to 0.10 s.

About 2200 strong-motion accelerographs were installed around the world as of 1974, and the number is said to have doubled shortly thereafter (Cherry, 1974). Borcherdt and Matthiesen (1980) show details of a strong-motion-instrumentation program in the United States. The record of a strong-motion accelerograph may include errors caused by factors such as (1) frequency characteristics of the accelerograph; (2) elongation and meandering of the recording paper; (3) offset, deviation, and friction of the recording pen; and (4) errors in reading the record, etc. These errors are usually corrected by filtering and other procedures.

1.2.3 Field Observation of Ground Motion

To provide a reliable basis for the seismic design of structures, the earthquake wave must be predicted as precisely as possible. Efforts have therefore been made to collect records of many strong earthquakes in many places. The shape and size of an earthquake wave depend not only on the source mechanism of the earthquake but also on travel-path geology, local site conditions, and other factors. Hence the record of a seismic wave taken at one place is sometimes modified before applying it to the prediction of the wave at another place.

The record of a strong earthquake usually consists of two horizontal north-south and east-west components and one vertical component. Figure 1-20 shows the east-west component of the ground acceleration recorded in the Taft earthquake (1952). The velocity and displacement of the ground can be obtained by integrating the acceleration record (see Fig. 1-28 below). In general, the two horizontal components are of comparable magnitude while the vertical component is usually somewhat smaller and includes more higher-frequency components.

The ground is always vibrating, with amplitudes as small as 10^{-6} to 10^{-7} m and with vibration periods ranging from several tenths of a second up to several seconds. Such ground vibration is referred to as a *microtremor*. Microtremors are caused by nearby traffic, machines in operation, and other moving objects. Their amplitude is usually larger in the daytime than at night, while the period is unchanged throughout the day.

Shown in Fig. 1-21*a* is the shape of a microtremor wave and in Fig. 1-21*b* its spectrum. Since the shape of a microtremor spectrum is similar to that of an actual earthquake, it has been suggested that microtremor records could be used to predict the seismic characteristics of strong earthquakes (Kanai, 1962; Kanai, 1969*b*); this technique has been put into practice. However, care must be taken in using microtremor records because (1) a microtremor is more sensitive to conditions at the vibration source, (2) it travels a different path, and (3) its amplitude is smaller than that of a strong earthquake. For these reasons, predictions based on microtremors can be misleading (Cherry, 1974).

1.2.4 Analysis of Earthquake Waves

One technique for determining ground-motion characteristics from a strong-motion-seismograph record is to construct a Fourier amplitude

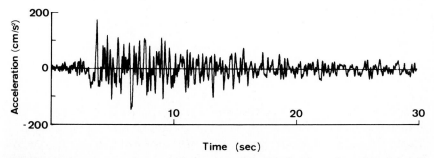

Figure 1-20 East-west component of ground acceleration for the Taft earthquake, California, 1952.

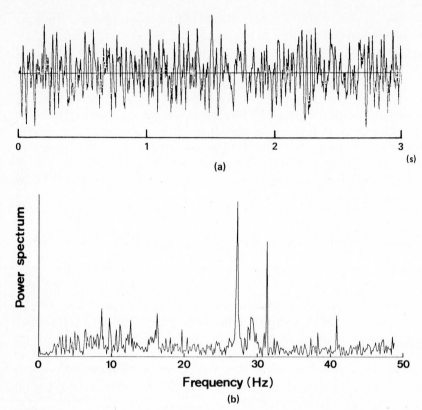

Figure 1-21 Data from microtremor observation. *(a)* Shape of waves. *(b)* Power spectrum.

spectrum by plotting the function $F(i\,\omega)$, given by Eq. (1-13), against the circular frequency ω (Fig. 1-22). Instead of the circular frequency, the ordinary frequency or the period is often used in the abscissa.

$$F(i\omega) = \int_{-\infty}^{\infty} f(t) \exp\left(-i\omega t\right) dt \qquad (1\text{-}13)$$

The concept of power spectral density is similar to that of the Fourier spectrum. Power spectral density expresses the energy content at each period and is proportional to the square of the Fourier amplitude spectrum.

For the purpose of structural studies, a response spectrum is widely used to represent ground-motion characteristics. The response spectrum is a plot of the maximum response to an earthquake of a single-degree-of-freedom system, with linear restoring-force characteristics, against the

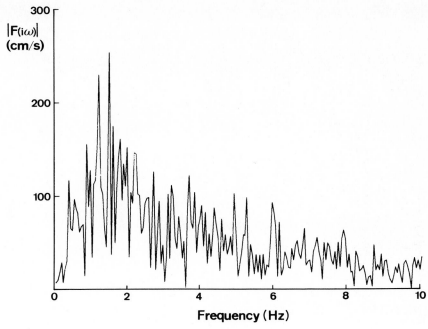

Figure 1-22   Fourier amplitude spectrum from the E1 Centro earthquake of 1940, north-south component.

natural period of the system. The response spectrum indicates the maximum amplitude of vibration that a structure will show to ground motion. The response spectrum will be referred to in detail in Sec. 2.1.5.

The spectrum intensity (SI) defined by Housner (1959) is also used frequently to express the effect of ground motion upon a structure. This is the area of a velocity response spectrum (see Sec. 2.1.5) within a prescribed period for a specific damping value (Fig. 1-23). The SI represents an average scale of ground-motion intensity with regard to influence upon an elastic system (Housner, 1970).

1.3   Earthquake Motion

1.3.1   Amplification Characteristics of Surface Layers

When seismic waves propagate in the ground, they reflect and refract at the borders of surface layers of the earth that have different qualities. If a seismic wave enters a softer layer, its path approaches the vertical axis more closely (see Fig. 1-24). That is, if c_1 and c_2 are the wave velocities of

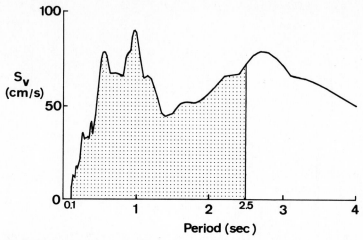

Figure 1-23　Spectrum intensity.

the lower and the upper layers respectively, then the following relation holds:

$$\frac{\sin \theta_1}{c_1} = \frac{\sin \theta_2}{c_2} \tag{1-14}$$

As illustrated in Fig. 1-25, this equation holds at each boundary surface. The relation between the angle of incidence θ_1 in the lowermost base rock and the angle of the wave θ_n in the top layer, irrespective of the properties of the intermediate layers, is

$$\sin \theta_n = \frac{c_n}{c_1} \sin \theta_1 \tag{1-15}$$

For example, if $c_n = 0.1c_1$ and $\theta_1 = 90°$, then $\theta_n \fallingdotseq 6°$. This indicates that the direction of wave propagation is almost vertical when the wave reaches the surface of the ground.

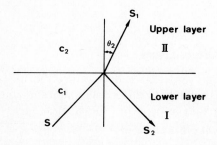

Figure 1-24　Reflection and re-fraction of waves.

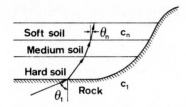

Figure 1-25 Reflection of waves in the surface layers.

As will be stated in Sec. 2.3.3, if a stationary wave enters from hard base rock, the amplitude of the wave at the surface is amplified and is greater than that of the original wave. The phenomenon of resonance takes place especially when the period of the wave coincides with the natural period of the surface layer. Since actual seismic waves are nonstationary waves, amplification is smaller than in stationary-wave propagation. Nevertheless, the seismic wave is gradually amplified as it propagates toward the surface of the ground, as shown in the numerical example of Fig. 1-26 (Toki, 1981).

1.3.2 Earthquake Motion on the Ground Surface

The shape, amplitude, duration, and other characteristics of a seismic wave are affected not only by the size of the earthquake and the hypocentral distance but also by the source mechanism, transmission-path geology, and local site conditions (Cherry, 1974). In other words, the shape of a seismic wave is altered by the change in the stress drop, the

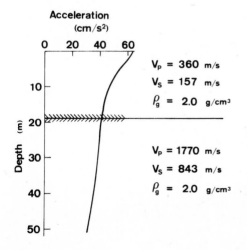

Figure 1-26 Distribution of root mean square of acceleration amplitude in the ground for the north-south component of the El Centro earthquake, California, 1940. (*From K. Toki,* Aseismic Analysis of Structures, *Gihodo Publishing Co., Tokyo, 1981.*)

maximum fault dislocation, and the area, shape, and nature of the fault surface. The wave shape at the focus is thus revealed at a point near the focus, but at a point on the far side of the focus not only is the intensity weakened but the wave shape is altered. That is, since short-period waves are more attenuated with wave transmission, the predominant period is apt to lengthen. The degree of amplification and the shape of a wave are affected also by the hardness and thickness of the layer under the local site.

Seismic waves are thus complex and different from one another. Newmark and Rosenblueth (1971) classified them into four types:

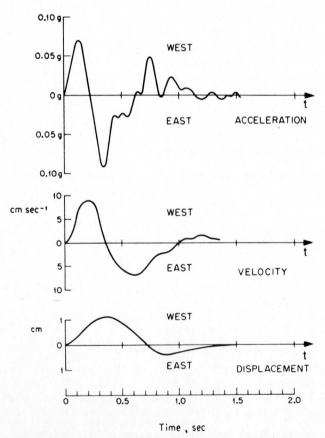

Figure 1-27 East-west component of the Port Hueneme earthquake, 1957. [*From G. W. Housner and D. E. Hudson, The Port Hueneme earthquake of March 18, 1957,* Bull. Seismol. Soc. Am., **48**(2), 163–168 (1958).]

1. *Single-shock type.* The focus is at a shallow depth, and the bedrock is hard, as in the Port Hueneme earthquake of 1957 (Fig. 1-27), the Libya earthquake of 1963, and the Skopje earthquake of 1963.

2. *A moderately long, extremely irregular motion.* The depth of the focus is intermediate, and the bedrock is hard, as in the El Centro earthquake of 1940. This type is often observed in the Circum-Pacific belt, where the bedrock is hard (Fig. 1-28).

3. *A long ground motion exhibiting pronounced prevailing periods of vibration.* The wave is filtered by many soft layers, and the successive reflections occur at the boundaries, as in the Mexican earthquake of 1964.

4. *A ground motion involving large-scale permanent deformation of the ground.* This occurred at Anchorage in the Alaskan earthquake of 1964 and in the Niigata earthquake of 1964.

As a matter of course, numerous earthquakes display wave shapes which are intermediate between or combine these four types.

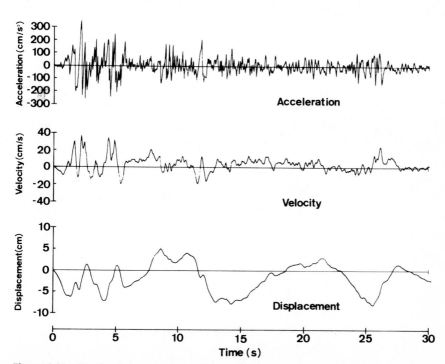

Figure 1-28   North-south component of the El Centro earthquake, California, 1940.

1.3.3 Relation between the Nature of the Ground and Structural Damage

Experience shows that structural damage is greater in soft alluvial ground formations for these reasons:

1. The seismic wave is amplified.

2. A structure with a certain natural period causes resonance phenomena as it collapses.

3. Uneven settlement, fissures, landslides, and other ground damage often occur.

Much structural damage on soft alluvial formations was reported after the old Nobi earthquake (1891), the San Francisco earthquake (1906), and the Kanto earthquake (1923) and after relatively recent earthquakes such as the Fukui earthquake (1948). In the Chile (1960), Peru (1970), Guatemala (1976), Vrancea (1977), and Miyagiken-Oki (1978) earthquakes structural damage was concentrated in regions where the ground was soft, and greater damage was found in rather remote city areas where the ground was soft than in regions closer to the epicenter.

REFERENCES

AIJ (Architectural Institute of Japan) (1981). *Data for Earthquake Resistant Design of Buildings*, AIJ, Tokyo (in Japanese).

Berlin, G. L. (1980). *Earthquakes and Urban Environment*, vol. 1, CRC Press, Inc., Boca Raton, Fla.

Bolt, B. A. (1978). *Earthquakes—A Primer*, W. H. Freeman and Company, San Francisco.

Borcherdt, R. D., and R. B. Matthiesen (1980). U.S. strong motion programs, *Proc. Seventh World Conf. Earthquake Eng.*, Istanbul, **2**, 9–16.

Cherry, S. (1974). Earthquake ground motions: Measurement and characteristics, in J. Solnes (ed.), *Engineering Seismology and Earthquake Engineering*, Noordhoff International Publishing, Leiden, 87–124.

Chinnery, M. A. (1969). Earthquake magnitude and source parameters, *Bull. Seismol. Soc. Am.*, **59**, 1969–1982.

Dowrick, D. J. (1977). *Earthquake Resistant Design*, John Wiley & Sons, Inc., New York.

Esteva, L., and E. Rosenblueth (1964). Espectros de temblores a distancias moderadas y grandes, *Bol. Soc. Mex. Ing. Sismol.*, **2**(1), 1–18.

Gutenberg, B., and C. F. Richter (1954). *Seismicity of the Earth and Associated Phenomena*, 2d ed., Princeton University Press, Princeton, N.J.

——— and ——— (1956). Earthquake magnitude, intensity, energy, and acceleration (second paper), *Bull. Seismol. Soc. Am.*, **46**(2), 105–145.

Halverson, H. T. (1965). The strong motion accelerograph, *Proc. Third World Conf. Earthquake Eng.*, Auckland, **1**, III 75–III 93.

Housner, G. W. (1959). Behavior of structures during earthquakes, *J. Eng. Mech. Div.*, *Am. Soc. Civ. Eng.*, **85**(EM-4), 109–129.

——— (1969). Engineering estimates of ground shaking and maximum earthquake magnitude, *Proc. Fourth World Conf. Earthquake Eng.*, Santiago, **1**(A-1), 1–13.

―――― (1970). Strong ground motion, in R. L. Wiegel (ed.), *Earthquake Engineering*, Prentice-Hall, Inc., Englewood Cliffs, N.J., 75−91.

―――― and D. E. Hudson (1958). The Port Hueneme earthquake of March 18, 1957, *Bull. Seismol. Soc. Am.*, **48**(2), 163−168.

Hudson, D. E. (1970). Ground motion measurements, in R. L. Wiegel (ed.), *Earthquake Engineering*, Prentice-Hall, Inc., Englewood Cliffs, N.J., 107−125.

Iida, K. (1965). Earthquake magnitude, earthquake fault, and source dimensions, *J. Earth Sci.*, Nagoya University, **13**, 115−132.

Kanai, K. (1962). On the spectrum of strong motion earthquakes, *Primeras journadas argentinas de ingenieria antisismica*, 24-1.

―――― (1969a). *Earthquake Engineering*, Kyoritsu Shuppan Co., Tokyo (in Japanese).

―――― (1969b). Earthquake motion and ground motion, in S. Aoki (ed.), *Earthquake Engineering*, Kyoritsu Shuppan Co., Tokyo, 9−42 (in Japanese).

Kasahara, K. (1981). *Earthquake Mechanics*, Cambridge University Press, London.

Ming-Tung, H. (1965). Seismicity of Taiwan, *Proc. Third World Conf. Earthquake Eng.*, Auckland, **1**, 3-116−124.

Newmark, N. M., and E. Rosenblueth (1971). *Fundamentals of Earthquake Engineering*, Prentice-Hall, Inc., Englewood Cliffs, N.J.

Nuttli, O. W. (1982). The earthquake problem in the Eastern United States, *J. Struct. Div.*, *Am. Soc. Civ. Eng.*, **108**(ST-6), 1302−1312.

Richter, C. F. (1935). An instrumental earthquake magnitude scale, *Bull. Seismol. Soc. Am.*, **25**, 1−32.

Tocher, D. (1958). Earthquake energy and ground breakage, *Bull. Seismol. Soc. Am.*, **48**, 147−153.

Toki, K. (1981). *Aseismic Analysis of Structures*, Gihodo Publishing Co., Tokyo (in Japanese).

Utsu, T. (1977). *Seismology*, Kyoritsu Shuppan Co., Tokyo (in Japanese).

2

VIBRATION OF STRUCTURES UNDER GROUND MOTION

2.1 Elastic Vibration of Simple Structures

2.1.1 Modeling of Structures and Equations of Motion

When a structure is analyzed for its vibrational characteristics, it first must be represented by a simple model which reflects its mechanical properties adequately. In many analyses, mass is assumed to be concentrated at the floor level of each story. By using this assumption, a single-story structure can be simplified as shown in Fig. 2-1. In this figure, the dashpot represents damping of the structure. The spring constant k relates the lateral deflection v to the shear force F_S.

Figure 2-1 illustrates a single-degree-of-freedom (SDOF) system subjected to a time-varying force $F(t)$. The viscous damping force F_D is the product of the damping coefficient c and the velocity. By applying d'Alembert's principle, the equation of motion is expressed as

$$F_I + F_D + F_S = F(t) \tag{2-1}$$

in which F_I = inertial force

$\quad\quad F_D$ = damping force

$\quad\quad F_S$ = force resisted by the spring

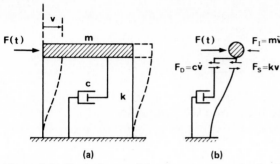

Figure 2-1 Single-degree-of-freedom system under horizontal force.

Then

$$m\ddot{v} + c\dot{v} + kv = F(t) \tag{2-2}$$

When the structure is subjected to a ground acceleration $\ddot{v}_g$ (Fig. 2-2), F_I can be expressed as

$$F_I = m(\ddot{v} + \ddot{v}_g)$$

Then

$$m\ddot{v} + c\dot{v} + kv = -m\ddot{v}_g \tag{2-3}$$

2.1.2 Free Vibration of Simple Structures

If no ground motion is applied to an SDOF system without damping, the governing equation of motion for free vibration, Eq. (2-3), can be simplified to

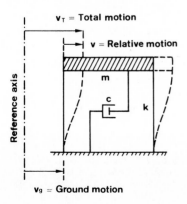

Figure 2-2 Single-degree-of-freedom system under ground motion.

$$m\ddot{v} + kv = 0 \tag{2-4}$$

The solution of Eq. (2-4) is given as

$$v = A \cos \omega t + B \sin \omega t \qquad \omega = \left(\frac{k}{m}\right)^{\frac{1}{2}} \tag{2-5}$$

Constants A and B can be determined from the initial conditions. $(A^2 + B^2)^{1/2}$ and ω, respectively, are the amplitude and circular frequency of the system. When $v = v(0)$ and $\dot{v} = \dot{v}(0)$ at time $t = 0$, the constants A and B are

$$A = v(0) \qquad B = \frac{\dot{v}(0)}{\omega}$$

Figure 2-3 illustrates the relation between deflection v and time t for the above condition. The natural period T is defined as the time required for the phase angle ωt to travel from zero to 2π. Then

$$\omega t = 2\pi$$

and

$$T = \frac{2\pi}{\omega} = 2\pi \left(\frac{m}{k}\right)^{\frac{1}{2}} \tag{2-6}$$

If viscous damping is present in the system, the equation of motion can be written as

$$m\ddot{v} + c\dot{v} + kv = 0 \tag{2-7}$$

or by dividing the equation by m,

$$\ddot{v} + 2\xi\omega\dot{v} + \omega^2 v = 0 \tag{2-8}$$

where

$$2\xi\omega = \frac{c}{m} \qquad \omega^2 = \frac{k}{m} \tag{2-9}$$

The solution of Eq. (2-8) is

$$v = A \exp(\lambda_1 t) + B \exp(\lambda_2 t) \tag{2-10}$$

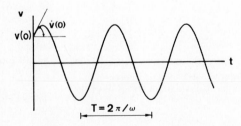

Figure 2-3 Undamped free-vibration response.

where $\quad \lambda_1, \ \lambda_2 = \omega[-\xi \pm (\xi^2 - 1)^{\frac{1}{2}}\ ]$ $\qquad(2\text{-}11)$

Equation (2-10) indicates that the solution changes its form according to the value of ξ.

If $\xi^2 < 1$,

$$v = \exp(-\xi\omega t)\ (A\ \cos\ \omega_D t + B\ \sin\ \omega_D t) \qquad(2\text{-}12)$$

or $\quad v = C \exp(-\xi\omega t) \sin(\omega_D t + \theta)$ $\qquad(2\text{-}13)$

where $\quad C = (A^2 + B^2)^{\frac{1}{2}} \qquad \theta = \tan^{-1} A/B$

$$\omega_D = (1 - \xi^2)^{\frac{1}{2}}\ \omega \qquad\qquad\qquad(2\text{-}14)$$

ω_D is called the *damped vibration frequency*. The envelope of the vibration amplitude is obtained as

$$C \exp(-\xi\omega t) = (A^2 + B^2)^{\frac{1}{2}} \exp(-\xi\omega t)$$

The system oscillates about the neutral position as the amplitude decays with time t.

If $\xi^2 > 1$, the system does not oscillate because the effect of damping overshadows the oscillation. This type of damping is therefore called *overdamping*.

The condition $\xi^2 = 1$ indicates a limiting value of damping at which the system loses its vibratory characteristics; this is called *critical damping*. If c_{cr} denotes the damping coefficient at critical damping, one can obtain from Eq. (2-9)

$$c_{cr} = 2\omega m = 2(mk)^{\frac{1}{2}} \qquad(2\text{-}15)$$

ξ is defined in terms of c_{cr} or as

$$\xi = \frac{c}{c_{cr}} \qquad(2\text{-}16)$$

ξ is the ratio of the coefficient of viscous damping to its value at critical damping and is called the *fraction of critical damping* or simply the *damping ratio*. Constants A, B, C, and θ in Eqs. (2-12) and (2-13) are determined from the initial conditions. As an example, if $v = 0$ and $\dot{v} = \dot{v}(0)$ at time $t = 0$,

$$v = \frac{\dot{v}(0)}{\omega_D} \exp(-\xi\omega t) \sin \omega_D t \qquad(2\text{-}17)$$

and the v versus t relationship can be drawn as in Fig. 2-4. The natural period T_D is given as

$$T_D = \frac{2\pi}{(1 - \xi^2)^{\frac{1}{2}}\ \omega} = \frac{T}{(1 - \xi^2)^{\frac{1}{2}}} \qquad(2\text{-}18)$$

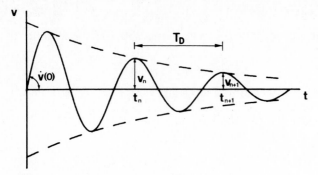

Figure 2-4   Free-vibration response of a damped system.

in which $T = (2\pi/\omega)$ is the natural period when the system is undamped. $T \doteq T_D$ if $\xi \ll 1$.

If the amplitudes at times t_n and $t_n + T_D$ are v_n and v_{n+1}, respectively, the ratio v_n/v_{n+1} can be given as

$$\frac{v_n}{v_{n+1}} = \exp \frac{2\pi\xi}{(1 - \xi^2)^{\frac{1}{2}}} \tag{2-19}$$

This ratio is called the *amplitude-decay ratio*. By taking the natural logarithm of both sides of Eq. (2-19), one obtains the logarithmic decrement

$$\ln \frac{v_n}{v_{n+1}} = 2\pi \frac{\xi}{(1 - \xi^2)^{\frac{1}{2}}} \doteq 2\pi\xi \tag{2-20}$$

2.1.3 Steady-State Forced Vibrations

If the system in Fig. 2-2 is subjected to sinusoidal ground motion such as

$$\ddot{v}_g = \alpha_0 \sin \omega' t \tag{2-21}$$

Eq. (2-3) gives

$$\ddot{v} + 2\xi\omega\dot{v} + \omega^2 v = -\alpha_0 \sin \omega' t \tag{2-22}$$

The solution of Eq. (2-22) is the sum of the complementary solution of Eq. (2-13) and the particular solution. Since the oscillation corresponding to the complementary solution rapidly decays as time elapses, the particular solution controls the vibration of the system at the steady state. This steady-state vibration is referred to as the *forced vibration*. The particular solution is

$$v = C_1 \sin \omega' t + C_2 \cos \omega' t \tag{2-23}$$

Then, by substituting Eq. (2-23) into Eq. (2-22),

$$v(t) = -\frac{\alpha_0}{\omega^2}[(1 - \beta^2)^2 + 4\xi^2\beta^2]^{-\frac{1}{2}} \sin(\omega't - \theta) \qquad (2\text{-}24)$$

where
$$\theta = \tan^{-1}\frac{2\xi\beta}{1 - \beta^2} \qquad (2\text{-}25)$$
$$\beta = \omega'/\omega$$

A static external force equal to the inertial force ($m\alpha_0$) makes the system deform by $m\alpha_0/k = \alpha_0/\omega^2$. This deflection, labeled v_{st}, is then

$$v_{st} = -\frac{\alpha_0}{\omega^2} \qquad (2\text{-}26)$$

According to Eqs. (2-24) and (2-26), the ratio of the resultant response amplitude to the static deflection v_{st}, called the *dynamic displacement-magnification factor D_d*, is

$$D_d = \left|\frac{v}{v_{st}}\right| = [(1 - \beta^2)^2 + 4\xi^2\beta^2]^{-\frac{1}{2}} \qquad (2\text{-}27)$$

Figure 2-5 shows the relationship between D_d and the frequency ratio β. As ω' approaches ω, the amplitude increases; this tendency is magnified for smaller values of ξ. The condition at which $\omega' = \omega$, i.e., $\beta = 1$, is called *resonance*.

From Eq. (2-3), absolute acceleration is given as

$$\ddot{v} + \ddot{v}_g = -(\omega^2 v + 2\xi\omega\dot{v}) \qquad (2\text{-}28)$$

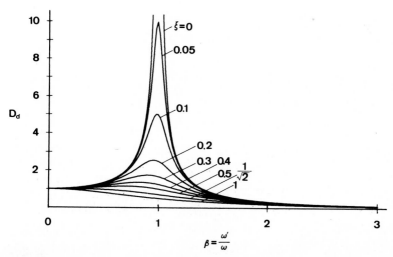

Figure 2-5 Dynamic displacement-magnification factor with damping and frequency as parameters.

By substituting Eq. (2-24) and its first derivative into the right side of Eq. (2-28),

$$\ddot{v} + \ddot{v}_g = \alpha_0[(1 - \beta^2)^2 + 4\xi^2\beta^2]^{-\frac{1}{2}}[1 + 4\xi^2\beta^2]^{\frac{1}{2}} \sin[\omega't - (\theta - \theta_0)]$$

$$(2\text{-}29)$$

where $\theta_0 = \tan^{-1} 2\xi\beta$ (2-30)

Thus, the ratio of the response to ground acceleration is

$$D_a = \left| \frac{\ddot{v} + \ddot{v}_g}{\ddot{v}_g} \right| = [(1 - \beta^2)^2 + 4\xi^2\beta^2]^{-\frac{1}{2}}[1 + 4\xi^2\beta^2]^{\frac{1}{2}} \qquad (2\text{-}31)$$

D_a is called the dynamic acceleration-magnification factor. The D_a versus β relationship is drawn in Fig. 2-6. At $\beta = \sqrt{2}$, the magnification factor D_a is unity regardless of the values of ξ, and D_a becomes smaller for smaller values of ξ in the range of $\beta > \sqrt{2}$.

2.1.4 Non-Steady-State Forced Vibrations

Discussion in Sec. 2.1.3 is devoted to the vibration of an SDOF system in its steady state. In the initial stage, the vibration corresponding to the complementary solution cannot be ignored. The vibration before reaching steady-state vibration is called *transient vibration*. For the sake of simplicity, it is assumed that $\omega' = \omega$ and that the ground motion $\ddot{v}_g =$

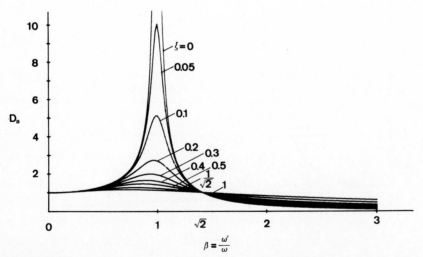

Figure 2-6 Dynamic acceleration-magnification factor with damping and frequency as parameters.

$\alpha_0 \sin \omega' t$; the solution is then obtained from Eqs. (2-12), (2-24), and (2-25) by setting $\beta = 1$:

$$v(t) = \exp{(-\xi\omega t)}(A \cos \omega_D t + B \sin \omega_D t) + (\alpha_0/\omega^2)[1(2\xi)] \cos \omega t \tag{2-32}$$

When $v = 0$ and $\dot{v} = 0$ at time $t = 0$, the constants A and B are

$$A = -\frac{\alpha_0}{\omega^2}\frac{1}{2\xi} \qquad B = \frac{\xi}{(1 - \xi^2)^{\frac{1}{2}}}A$$

Figure 2-7 schematically shows the fluctuation in amplitude, which gradually increases as time passes and approaches the amplitude in the steady state.

Let us consider the response of a damped SDOF system subjected to arbitrary ground motion. To compute the response, ground motion is assumed to correspond to the sum of a series of impulsive loads. The effective external force $F(t)$ caused by arbitrary ground motion is (Fig. 2-8):

$$F(t) = - m\ddot{v}_g \tag{2-33}$$

Taking $F(t)$ as an impulsive load applied during an infinitesimal time interval $d\tau$, and from the condition that the momentum $m\dot{v}$ equals the impulse $F(\tau)\,d\tau$, we can obtain

$$m\dot{v} = F(\tau)\,d\tau \tag{2-34}$$

This expression means that during a time change $d\tau$ the impulse makes the velocity of the mass change by $F(\tau)\,d\tau/m$. Therefore, the solution is obtained from the following initial conditions:

$$v = 0 \qquad \text{and} \qquad \dot{v} = F(\tau)\,d\tau/m \qquad \text{at} \qquad t = \tau$$

The initial conditions shown above express the solution of free vibration. By substituting the initial velocity $\dot{v}(0) = [F(\tau)/m]\,d\tau = -\ddot{v}_g(\tau)d\tau$ at $t = \tau$ and $t = t - \tau$ into Eq. (2-17), we obtain the following expression:

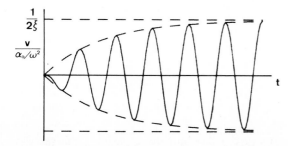

Figure 2-7 Transient response of a damped system.

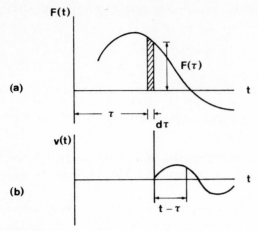

Figure 2-8 Derivation of the Duhamel integral.

$$v(t) = - [\ddot{v}_g(\tau) \, d\tau/\omega_D] \exp [-\xi\omega(t - \tau)] \sin \omega_D(t - \tau) \tag{2-35}$$

Equation (2-35) represents the vibration of the system when it is subjected to an impulsive load of $F(\tau) = -m\ddot{v}_g(\tau)$. When $F(t)$ is applied to the system continuously, the vibration of the system is obtained by summing Eq. (2-35) with respect to time τ. Thus

$$v(t) = - \frac{1}{\omega_D} \int_0^t \ddot{v}_g(\tau) \exp [-\xi\omega(t - \tau)] \sin \omega_D(t - \tau) \, d\tau \tag{2-36}$$

This equation is called the *Duhamel integral*. Since $\xi \ll 1$ in most building structures, $(1 - \xi^2)^{1/2} \doteq 1.0$ and Eq. (2-36) can be approximated as

$$v(t) \doteq - \frac{1}{\omega} \int_0^t \ddot{v}_g(\tau) \exp [-\xi\omega(t - \tau)] \sin \omega(t - \tau) \, d\tau \tag{2-37}$$

The velocity of the system is obtained by first differentiating Eq. (2-17) once and then using the procedure obtained in the derivation of Eq. (2-36). Thus

$$\dot{v}(t) \doteq - \int_0^t \ddot{v}_g(\tau) \exp [-\xi\omega(t - \tau)] \cos [\omega(t - \tau) + \psi] \, d\tau \tag{2-38}$$

where

$$\psi = \tan^{-1} \frac{\xi}{(1 - \xi^2)^{\frac{1}{2}}} \tag{2-39}$$

Analogously, (absolute) acceleration is given by neglecting the second term on the right side of Eq. (2-28) as

$$\ddot{v} + \ddot{v}_g \doteqdot \omega \int_0^t \ddot{v}_g(\tau) \exp\left[-\xi\omega(t - \tau)\right] \sin \omega(t - \tau) \, d\tau \tag{2-40}$$

2.1.5 Response-Spectrum Representation

In designing structures subjected to non-steady-state vibration, the major concern is the maximum (in the absolute sense) value of the response. The relative displacement v reaches its maximum when in Eq. (2-37) the integral takes the maximum value. With the maximum value of this quantity defined as S_v, we have

$$S_d = \frac{1}{\omega} S_v = \frac{T}{2\pi} S_v = v_{\max} \tag{2-41}$$

S_d is called the *spectral displacement,* and hence

$$S_v = \{\textstyle\int \ddot{v}_g(\tau) \exp\left[-\xi\omega(t - \tau)\right] \sin \omega(t - \tau) \, d\tau\}_{\max} \tag{2-42}$$

In structures with damping, S_v is not identical with, but is very close to, the maximum-velocity response. Therefore, S_v is considered the maximum velocity and is called the *spectral pseudo velocity* or simply the *spectral velocity.*

Then as a reasonable approximation

$$S_v \doteqdot \dot{v}_{\max} \tag{2-43}$$

According to Eqs. (2-40) and (2-42)

$$S_a = \omega S_v = \frac{2\pi}{T} S_v \doteqdot (\ddot{v} + \ddot{v}_g)_{\max} \tag{2-44}$$

S_a is called the *spectral acceleration* (or, more accurately, the *spectral pseudo acceleration* because S_a does not exactly represent the peak acceleration value in most cases).

The earthquake load applied to the structure, i.e., the maximum base shear $V_{\max}$, is

$$V_{\max} = mS_a \tag{2-45}$$

This equation indicates that the maximum base shear can readily be computed once the mass of a structure and the spectral acceleration are known. By using Eqs. (2-41) through (2-44), S_a, S_d, and S_v for an SDOF system subjected to an earthquake motion can be drawn with respect to each particular combination of natural period and damping coefficient. The diagrams plotting S_a, S_v, and S_d are called respectively the *acceleration response spectrum,* the *velocity response spectrum,* and the *displacement response spectrum.* Figure 2-9 plots the S_a of an SDOF system under the north-south component of the El Centro earthquake ground motion recorded in 1940. Once the natural period and the damping coefficient of a structure are known, the maximum response of the structure subjected to this earthquake motion can be found from this diagram. Further, the maximum base shear applied to the structure can be computed by the use of Eq. (2-45).

As shown in Fig. 2-9, response spectra vary greatly with the natural period. However, for design purposes generalized spectra rather than a specific spectrum are more meaningful. Figures 2-10 through 2-12 show the average response spectra devised by Housner (1959; 1970) from two components each of four different earthquake records in the United States. Similar spectra were proposed by Umemura (1962) on the basis of earthquake motions recorded in the United States and Japan.

As shown in Fig. 2-10, the velocity spectrum is nearly constant in a range of longer natural periods. The acceleration spectrum decreases as the natural period lengthens (Fig. 2-11), whereas the displacement spectrum increases in proportion to the natural period (Fig. 2-12). Rough sketches of these three spectra are shown in Fig. 2-13. As indicated by this figure, S_a is proportional to $1/T$ in the range of longer natural periods. With reference to Eq. (2-45), one can presume that the base shear is also proportional to $1/T$. In many seismic building codes, the base shear of a high-rise building is taken to be proportional to $1/T^{2/3}$ or $1/T^{1/2}$ rather than $1/T$ in order to provide a safety margin (see Table 4-1 in Chap. 4).

As Eqs. (2-41), (2-43), and (2-44) correlate S_a, S_v, and S_d with one

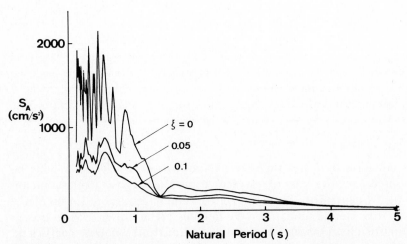

Figure 2-9 Acceleration response spectra derived from the north-south component of the El Centro earthquake, California, 1940.

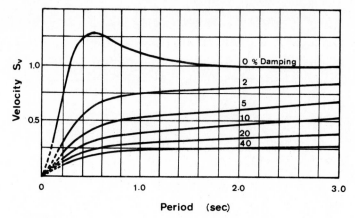

Figure 2-10 Smoothed average velocity response spectra S_v by Housner; arbitrary scale. [*G. W. Housner, Behavior of structures during earthquakes,* J. Eng. Mech. Div., Am. Soc. Civ. Eng., *85(EM-4), 109−129 (1959).*]

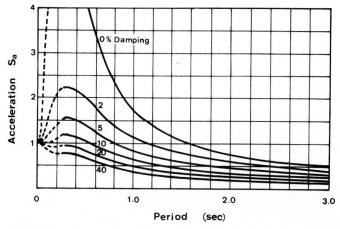

Figure 2-11 Smoothed average acceleration response spectra S_a by Housner; arbitrary scale. [*G. W. Housner, Behavior of structures during earthquakes,* J. Eng. Mech. Div., Am. Soc. Civ. Eng., *85(EM-4), 109−129 (1959).*]

another, those three spectra can be drawn in one figure as shown in Fig. 2-14. In this figure, the abscissa denotes the natural period and the ordinate the spectrum velocity, with both of those axes following a logarithmic scale. S_d and S_a are read from axes which are inclined at −45° and +45° to the abscissa.

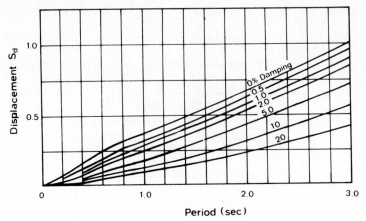

Figure 2-12 Smoothed average displacement spectra S_d by Housner; arbitrary scale. (*G. W. Housner, Strong ground motion*, in *R. L. Wiegel (ed.)*, Earthquake Engineering, *Prentice-Hall, Inc., Englewood Cliffs, N.J., 1970, 75−91.*)

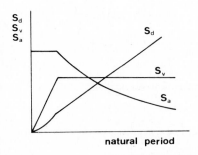

Figure 2-13 General shapes of response spectra.

2.2   Elastic Vibration of Multistory Structures

2.2.1   Equations of Motion

In analyzing a multistory structure subjected to time-varying forces, a system model with masses and springs as shown in Fig. 2-15 is frequently used. Roughly speaking, multistory structures can be divided into two groups according to their deformation characteristics: in one group the floors move only in the horizontal direction, and these are referred to as shear-type structures; in the other group the floors move in both rotational and horizontal directions, and they are referred to as moment-shear-type structures.

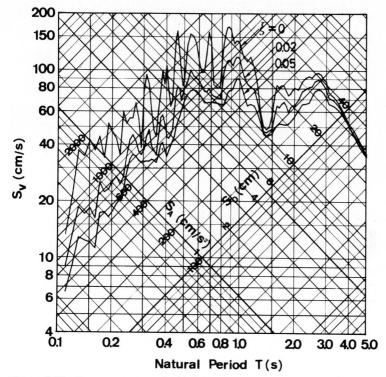

Figure 2-14 Response spectra derived from north-south component of the El Centro earthquake, California, 1940.

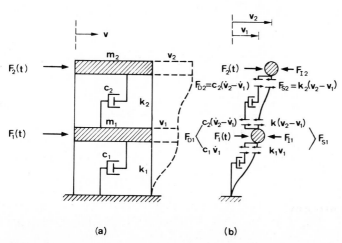

Figure 2-15 Two-degrees-of-freedom system under horizontal forces.

The equations of motion of the system in Fig. 2-15 are

$$F_{I1} + F_{D1} + F_{S1} = F_1(t)$$
$$F_{I2} + F_{D2} + F_{S2} = F_2(t) \tag{2-46}$$

The inertial forces in the equations are

$$F_{I1} = m_1 \ddot{v}_1$$
$$F_{I2} = m_2 \ddot{v}_2 \tag{2-47}$$

or in matrix form,

$$\begin{Bmatrix} F_{I1} \\ F_{I2} \end{Bmatrix} = \begin{bmatrix} m_1 & 0 \\ 0 & m_2 \end{bmatrix} \begin{Bmatrix} \ddot{v}_1 \\ \ddot{v}_2 \end{Bmatrix} \tag{2-48}$$

or $$\mathbf{F}_I = \mathbf{m}\ddot{\mathbf{v}} \tag{2-49}$$

Here $\mathbf{F}_I$, $\ddot{\mathbf{v}}$, and $\mathbf{m}$ are the inertial-force vector, acceleration vector, and mass matrix, respectively. As shown in Fig. 2-15b, the lumped masses are concentrated at floor levels, and the mass matrix is therefore a diagonal matrix. The restoring forces and displacements are related as follows:

$$F_{S1} = k_1 v_1 - k_2(v_2 - v_1)$$
$$F_{S2} = k_2(v_2 - v_1) \tag{2-50}$$

By introducing k_{11}, k_{12}, k_{21}, k_{22},

$$k_{11} = k_1 + k_2 \quad \text{and} \quad k_{12} = -k_2$$
$$k_{21} = -k_2 \quad \text{and} \quad k_{22} = k_2 \tag{2-51}$$

and substituting into Eq. (2-50), the following equations are obtained (Fig. 2-16):

$$F_{S1} = k_{11}v_1 + k_{12}v_2$$
$$F_{S2} = k_{21}v_1 + k_{22}v_2 \tag{2-52}$$

k_{ij} can be explained as the force applied to the ith story when the jth story is subjected to a unit displacement while all other stories remain undisplaced. According to the Maxwell-Betti reciprocal theorem,

$$k_{ij} = k_{ji} \tag{2-53}$$

A matrix expression for Eq. (2-52) is

$$\begin{Bmatrix} F_{S1} \\ F_{S2} \end{Bmatrix} = \begin{bmatrix} k_{11} & k_{12} \\ k_{21} & k_{22} \end{bmatrix} \begin{Bmatrix} v_1 \\ v_2 \end{Bmatrix} \tag{2-54}$$

or $$\mathbf{F}_S = \mathbf{k}\mathbf{v} \tag{2-55}$$

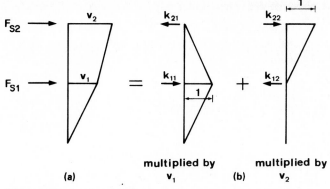

multiplied by   multiplied by

(a) v_1 (b) v_2

Figure 2-16   Load and deflection of a two-degrees-of-freedom system. (a) Total deflection. (b) Decomposition of deflection.

Here $\mathbf{F}_S$, $\mathbf{v}$, and $\mathbf{k}$ are the elastic-force vector, displacement vector, and stiffness matrix, respectively. Equation (2-53) indicates that $\mathbf{k}$ is a symmetrical matrix.

If damping forces induced by viscous damping are assumed to be proportional to relative velocities,

$$\begin{Bmatrix} F_{D1} \\ F_{D2} \end{Bmatrix} = \begin{bmatrix} c_{11} & c_{12} \\ c_{21} & c_{22} \end{bmatrix} \begin{Bmatrix} \dot{v}_1 \\ \dot{v}_2 \end{Bmatrix} \tag{2-56}$$

or in matrix form,

$$\mathbf{F}_D = \mathbf{c}\dot{\mathbf{v}} \tag{2-57}$$

Here $\mathbf{F}_D$, $\dot{\mathbf{v}}$, and $\mathbf{c}$ are the viscous damping-force vector, velocity vector, and viscous damping matrix, respectively. The applied load vector is

$$\begin{Bmatrix} F_1(t) \\ F_2(t) \end{Bmatrix} \qquad \text{or} \qquad \mathbf{F}(t) \tag{2-58}$$

By using Eqs. (2-49), (2-55), (2-57), and (2-58), the equations of motion for the two-degrees-of-freedom system can be written as

$$\mathbf{F}_I + \mathbf{F}_D + \mathbf{F}_S = \mathbf{F}(t) \tag{2-59}$$

or $\quad$ $$\mathbf{m}\ddot{\mathbf{v}} + \mathbf{c}\dot{\mathbf{v}} + \mathbf{k}\mathbf{v} = \mathbf{F}(t) \tag{2-60}$$

This expression is essentially in the same form as the equation for an SDOF system [Eq. (2-2)].

If ground acceleration $\ddot{v}_g$ is applied to the structure, then

$$\mathbf{m}\ddot{\mathbf{v}} + \mathbf{c}\dot{\mathbf{v}} + \mathbf{k}\mathbf{v} = -\mathbf{m}\mathbf{1}\ddot{v}_g \tag{2-61}$$

in which $\mathbf{1}$ is a unit vector. This equation and Eq. (2-3), the equation for an SDOF system, are also intrinsically the same.

2.2.2 Periods and Modes of Vibration of Structural Systems

For an undamped multiple-degrees-of-freedom (MDOF) system in free vibration, Eq. (2-61) becomes

$$\mathbf{m}\ddot{\mathbf{v}} + \mathbf{k}\mathbf{v} = 0 \tag{2-62}$$

The solution of the equation is assumed to be

$$\mathbf{v} = \hat{\mathbf{v}} \sin \omega t \tag{2-63}$$

$\hat{\mathbf{v}}$ represents the vibrational shape of the system. By differentiating Eq. (2-63) twice,

$$\ddot{\mathbf{v}} = - \omega^2 \hat{\mathbf{v}} \sin \omega t \tag{2-64}$$

Substituting Eqs. (2-63) and (2-64) into Eq. (2-62) gives the expression

$$\mathbf{k}\hat{\mathbf{v}} - \omega^2 \mathbf{m}\hat{\mathbf{v}} = 0 \tag{2-65}$$

Equation (2-65) is called the frequency equation with respect to the circular frequency ω. When the system has n degrees of freedom, n natural circular frequencies are obtained from Eq. (2-65). The lowest value of ω is called the first natural circular frequency ω_1. The ω are numbered sequentially so that the nth lowest value of ω is the nth natural circular frequency; by substituting it into Eq. (2-65), relative displacements $\hat{\mathbf{v}}$ of the system which represent the shape of vibration, called the *modal shape,* can be determined. For a two-degrees-of-freedom system, Eq. (2-65) is

$$\begin{aligned} (k_{11} - \omega^2 m_1)\hat{v}_1 + k_{12}\hat{v}_2 = 0 \\ k_{21}\hat{v}_1 + (k_{22} - \omega^2 m_2) \hat{v}_2 = 0 \end{aligned} \tag{2-66}$$

For $\hat{\mathbf{v}}$ to have a nontrivial solution, the determinant of Eq. (2-66) must be zero:

$$\begin{vmatrix} k_{11} - \omega^2 m_1 & k_{12} \\ k_{21} & k_{22} - \omega^2 m_2 \end{vmatrix} = 0 \tag{2-67}$$

That is,

$$(m_1\omega^2 - k_{11})(m_2\omega^2 - k_{22}) - k_{12}k_{21} = 0 \tag{2-68}$$

Among the four roots derived from Eq. (2-68), the positive roots ω_1 and ω_2 respectively correspond to the first and second natural circular frequencies. By substituting them into Eq. (2-66), the ratio of displacements $\hat{v}_2/\hat{v}_1$ is uniquely determined for each ω_1 and ω_2 as shown in Fig. 2-17. The modal shapes corresponding to ω_1 and ω_2 are called the first and second mode, respectively. As evidenced from the condition specified by Eq. (2-66), only displacement ratios of $\hat{v}$ can be obtained. In usual practice,

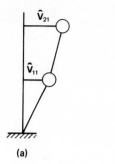
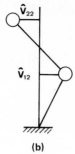

(a)   **(b)**

Figure 2-17 Modal shapes of a two-degrees-of-freedom system. (*a*) First mode. (*b*) Second mode.

the displacement corresponding to the top or lowest story for the maximum displacement is taken to be unity as a datum. If a system has N degrees of freedom, the nth modal shape $\boldsymbol{\phi}_n$ is written as

$$\boldsymbol{\phi}_n = \begin{bmatrix} \phi_{1n} \\ \phi_{2n} \\ \cdot \\ \cdot \\ \cdot \\ \phi_{Nn} \end{bmatrix} \equiv \frac{1}{\hat{v}_{kn}} \begin{bmatrix} \hat{v}_{1n} \\ \hat{v}_{2n} \\ \cdot \\ \cdot \\ \cdot \\ \hat{v}_{Nn} \end{bmatrix} \tag{2-69}$$

Here $\hat{v}$ represents the reference component. The square matrix, consisting of n-modal-shape vectors, is called the modal-shape matrix and is expressed as

$$\boldsymbol{\phi} = [\boldsymbol{\phi}_1 \boldsymbol{\phi}_2 \cdots \boldsymbol{\phi}_N] = \begin{bmatrix} \phi_{11} & \phi_{12} & \cdots & \phi_{1N} \\ \phi_{21} & \phi_{22} & \cdots & \phi_{2N} \\ \cdot & \cdots & \cdots & \cdot \\ \phi_{N1} & \phi_{N2} & \cdots & \phi_{NN} \end{bmatrix} \tag{2-70}$$

2.2.3 Orthogonality of Vibration Modes

Modal-shape vectors possess an orthogonality relationship. To demonstrate this relationship, let us consider $\hat{v}_n$, the nth-modal-shape vector. From Eq. (2-65), one can obtain

$$\mathbf{k}\hat{\mathbf{v}}_n - \omega_n^2 \mathbf{m}\hat{\mathbf{v}}_n = \mathbf{0} \tag{2-71}$$

By premultiplying Eq. (2-71) by the transpose of the mth-modal-shape vector $\hat{\mathbf{v}}_m$, one obtains

$$\hat{\mathbf{v}}_m^T \mathbf{k} \hat{\mathbf{v}}_n - \omega_n^2 \hat{\mathbf{v}}_m^T \mathbf{m} \hat{\mathbf{v}}_n = 0 \tag{2-72}$$

Interchanging m and n in Eq. (2.72) gives

$$\hat{\mathbf{v}}_n^T \mathbf{k} \, \hat{\mathbf{v}}_m - \omega_m^2 \hat{\mathbf{v}}_n^T \mathbf{m} \hat{\mathbf{v}}_m = 0 \tag{2-73}$$

Considering the symmetrical characteristics of matrices $\mathbf{m}$ and $\mathbf{k}$,

$$\hat{\mathbf{v}}_m^T \mathbf{k} \, \hat{\mathbf{v}}_n = \hat{\mathbf{v}}_n^T \mathbf{k} \, \hat{\mathbf{v}}_m$$
$$\hat{\mathbf{v}}_m^T \mathbf{m} \, \hat{\mathbf{v}}_n = \hat{\mathbf{v}}_n^T \mathbf{m} \, \hat{\mathbf{v}}_m$$

and then subtracting Eq. (2-72) from Eq. (2-73) leads to

$$(\omega_n^2 - \omega_m^2) \, \hat{\mathbf{v}}_n^T \mathbf{m} \, \hat{\mathbf{v}}_m = 0$$

With the condition that $\omega_n^2 - \omega_m^2 \neq 0 \; (m \neq n)$, then

$$\hat{\mathbf{v}}_n^T \mathbf{m} \, \hat{\mathbf{v}}_m = 0 \qquad \text{with} \qquad m \neq n \tag{2-74}$$

Another expression of Eq. (2-74) is

$$\Sigma m_i \, \hat{v}_{in} \hat{v}_{im} = 0 \tag{2-75}$$

This equation indicates that the two modal-shape vectors $\hat{\mathbf{v}}_n$ and $\hat{\mathbf{v}}_m$ are orthogonal with respect to the mass matrix $\mathbf{m}$. Substituting Eq. (2-74) into Eq. (2-73) gives

$$\hat{\mathbf{v}}_n^T \mathbf{k} \, \hat{\mathbf{v}}_m = 0 \qquad \text{with} \qquad n \neq m \tag{2-76}$$

Thus, the modal-shape vectors also are orthogonal to each other with respect to the stiffness matrix $\mathbf{k}$. An N-degrees-of-freedom system contains N individual modal shapes. Arbitrary displacements $\mathbf{v}$ of the system can be expressed as the sum of the nth-modal-shape vectors $\boldsymbol{\phi}_n$ multiplied by an amplitude Y_n (Fig. 2-18):

$$\mathbf{v} = \sum_{n=1}^{N} \boldsymbol{\phi}_n Y_n \tag{2-77}$$

or by using the modal-shape matrix defined in Eq. (2-70),

$$\mathbf{v} = \boldsymbol{\phi} \mathbf{Y} \tag{2-78}$$

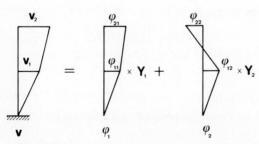

Figure 2-18 Deflections as the sum of modal components.

The vector **Y** is called the *general coordinate vector* or the *normal coordinates of the system*. By premultiplying Eq. (2-78) by $\boldsymbol{\phi}_n^T \mathbf{m}$ and considering the orthogonality condition indicated by Eq. (2-74), the amplitude corresponding to the nth modal shape Y_n can be derived:

$$\boldsymbol{\phi}_n^T \mathbf{m} \mathbf{v} = \boldsymbol{\phi}_n^T \mathbf{m} \boldsymbol{\phi}_n Y_n$$

$$Y_n = \frac{\boldsymbol{\phi}_n^T \mathbf{m} \mathbf{v}}{\boldsymbol{\phi}_n^T \mathbf{m} \boldsymbol{\phi}_n} \tag{2-79}$$

or

$$Y_n = \frac{\displaystyle\sum_{i=1}^{N} m_i \phi_{in} v_i}{\displaystyle\sum_{i=1}^{N} m_i \phi_{in}^2} \tag{2-80}$$

In the case of a two-degrees-of-freedom system,

$$Y_1 = \frac{m_1 \phi_{11} v_1 + m_2 \phi_{21} v_2}{m_1 \phi_{11}^2 + m_2 \phi_{21}^2} \qquad Y_2 = \frac{m_1 \phi_{12} v_1 + m_2 \phi_{22} v_2}{m_1 \phi_{12}^2 + m_2 \phi_{22}^2} \tag{2-81}$$

$\dot{\mathbf{v}}$ and $\ddot{\mathbf{v}}$ can also be expressed by using the normal coordinates since $\mathbf{v}$ is now expressed as in Eq. (2-77). When the equation for forced vibration [Eq. (2-61)] is to be solved with respect to the normal coordinates, the right side of the equation also must be expressed with respect to these coordinates. First, a unit vector **1** is decomposed to

$$\mathbf{1} = \sum_{n=1}^{N} \boldsymbol{\phi}_n \beta_n \tag{2-82}$$

or

$$\mathbf{1} = \boldsymbol{\phi} \boldsymbol{\beta} \tag{2-83}$$

In the case of the two-degrees-of-freedom system, the expression of Eq. (2-83) is as represented in Fig. 2-19:

$$\mathbf{1} = \boldsymbol{\phi}_1 \beta_1 + \boldsymbol{\phi}_2 \beta_2 \tag{2-84}$$

To find β_n, Eq. (2-83) is premultiplied by $\boldsymbol{\phi}_n^T \mathbf{m}$. Thus,

$$\boldsymbol{\phi}_n^T \mathbf{m} \mathbf{1} = \boldsymbol{\phi}_n^T \mathbf{m} \boldsymbol{\phi} \boldsymbol{\beta} \tag{2-85}$$

From the orthogonality condition,

$$\beta_n = \frac{\boldsymbol{\phi}_n^T \mathbf{m} \mathbf{1}}{\boldsymbol{\phi}_n^T \mathbf{m} \boldsymbol{\phi}_n} = \frac{\displaystyle\sum_{i=1}^{N} m_i \phi_{in}}{\displaystyle\sum_{i=1}^{N} m_i \phi_{in}^2} \tag{2-86}$$

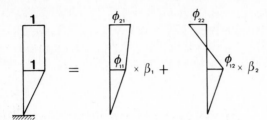

Figure 2-19 Unit deflection as the sum of modal components.

β_n represents the relative participation of the nth modal shape in the entire vibration of the system. It is called the *earthquake-participation factor* for the nth mode.

2.2.4 Modal-Analysis Technique

Equations of motion for an n-degrees-of-freedom system can be converted to n independent equations by the use of the normal coordinates. The equation of motion of a damped MDOF system subjected to free vibration is given by Eq. (2-60) with its right side set to zero:

$$\mathbf{m\ddot{v}} + \mathbf{c\dot{v}} + \mathbf{kv} = 0 \tag{2-87}$$

Substituting Eq. (2-78) into the above equation leads to

$$\mathbf{m\phi\ddot{Y}} + \mathbf{c\phi\dot{Y}} + \mathbf{k\phi Y} = 0 \tag{2-88}$$

Premultiplying Eq. (2-88) by $\mathbf{\phi}^T$ gives

$$\mathbf{\phi}^T\mathbf{m\phi\ddot{Y}} + \mathbf{\phi}^T\mathbf{c\phi\dot{Y}} + \mathbf{\phi}^T\mathbf{k\phi Y} = 0 \tag{2-89}$$

From the orthogonality condition, Eqs. (2-74) and (2-76) are still valid. If the orthogonality condition is assumed to be applicable to the damping matrix $\mathbf{c}$, all of $\mathbf{\phi}^T\mathbf{m\phi}$, $\mathbf{\phi}^T\mathbf{k\phi}$, and $\mathbf{\phi}^T\mathbf{c\phi}$ become diagonal matrices. Therefore, Eq. (2-89) is uncoupled to give

$$\mathbf{\phi}_n^T\mathbf{m\phi}_n\ddot{Y}_n + \mathbf{\phi}_n^T\mathbf{c\phi}_n\dot{Y}_n + \mathbf{\phi}_n^T\mathbf{k\phi}_nY_n = 0 \tag{2-90}$$

or $M_n\ddot{Y}_n + C_n\dot{Y}_n + K_nY_n = 0$

$$M_n = \mathbf{\phi}_n^T\mathbf{m\phi}_n \tag{2-91}$$
$$K_n = \mathbf{\phi}_n^T\mathbf{k\phi}_n$$
$$C_n = \mathbf{\phi}_n^T\mathbf{c\phi}_n$$

M_n, K_n, and C_n are called the generalized mass, generalized spring constant, and generalized damping coefficient of the nth mode, respectively. Equations (2-91) can also be written as

$$\ddot{Y}_n + 2\xi_n\omega_n\dot{Y}_n + \omega_n^2 Y_n = 0$$

$$\omega_n^2 = \frac{K_n}{M_n} \qquad \xi_n = \frac{C_n}{2M_n\omega_n} \tag{2-92}$$

Eventually, the equations of motion of the n-degrees-of-freedom system are decomposed to give n independent equations with respect to the normal coordinates. Since Eq. (2-92) is identical with the equation of motion of an SDOF system, solving an n-degrees-of-freedom system in free vibration is equivalent to first solving n independent SDOF systems in free vibration and then combining them according to their participation. This technique for solving an MDOF system is called *modal analysis.*

In analyzing an MDOF system subjected to ground motion, this technique is more effective than direct integration of Eq. (2-61). Substituting Eqs. (2-78) and (2-83) into Eq. (2-61) and applying the procedure for analyzing an MDOF system subjected to free vibration lead to

$$\ddot{Y}_n + 2\xi_n\omega_n\dot{Y}_n + \omega_n^2 Y_n = -\beta_n\ddot{v}_g \tag{2-93}$$

By setting

$$Y_n = \beta_n Y_{no} \tag{2-94}$$

Eq. (2-93) becomes

$$\ddot{Y}_{no} + 2\xi_n\omega_n\dot{Y}_{no} + \omega_n^2 Y_{no} = -\ddot{v}_g \tag{2-95}$$

This equation is identical with the equation for an SDOF system subjected to ground motion. According to Eq. (2-77), the displacements can be expressed as

$$\mathbf{v} = \sum_{n=1}^{N} \boldsymbol{\phi}_n \beta_n Y_{no} \tag{2-96}$$

Here $\beta_n\boldsymbol{\phi}_n$ is called the *earthquake-participation function.* The velocities are

$$\dot{\mathbf{v}} = \sum_{n=1}^{N} \boldsymbol{\phi}_n \beta_n \dot{Y}_{no} \tag{2-97}$$

The relative accelerations are

$$\ddot{\mathbf{v}} = \sum_{n=1}^{N} \boldsymbol{\phi}_n \beta_n \ddot{Y}_{no} \tag{2-98}$$

By decomposing $\ddot{v}_g$ into a vector form by the use of Eq. (2-82) and superposing it onto Eq. (2-98), one can obtain

$$\ddot{\mathbf{v}} + \mathbf{1}\ddot{v}_g = \sum_{n=1}^{N} \boldsymbol{\phi}_n \beta_n(\ddot{Y}_{no} + \ddot{v}_g) \tag{2-99}$$

The second term on the left side of Eq. (2-95) is much smaller than the first term and may be neglected:

$$\ddot{Y}_{no} + \ddot{v}_g = - \omega_n^2 Y_{no} \tag{2-100}$$

Substituting Eq. (2-100) into Eq. (2-99) leads to

$$\ddot{v} + 1\ddot{v}_g = - \sum_{n=1}^{N} \phi_n \beta_n \omega_n^2 Y_{no} \tag{2-101}$$

The solution of Eq. (2-95) for non-steady-state ground motion can be derived from Eq. (2-36):

$$Y_{no} = - \frac{1}{\omega_{Dn}} \int_0^t \ddot{v}_g(\tau) \exp\left[-\xi\omega_n(t - \tau)\right] \sin \omega_{Dn}(t - \tau) \, d\tau \tag{2-102}$$

As described above, forced vibration of the n-degrees-of-freedom system can be expressed as the sum of the forced vibrations of n independent SDOF systems.

In most engineering problems, only extreme values of displacements, velocities, and accelerations are needed. If the maximum values of all modes occur at the same instant, the maximum response is given as

$$v_{\text{max}} = |v_1|_{\text{max}} + |v_2|_{\text{max}} + \cdots$$

The extreme values, however, are unlikely to occur simultaneously; also, their signs cannot be the same. The best approach for predicting maximum response is to combine probabilistically the extreme values of all modes. Among many proposals for the probabilistic evaluation of maximum response, the most popular is the root-sum-square method. In this method, maximum displacement is expressed as

$$v_{\text{max}} = (v_{1_{\text{max}}}^2 + v_{2_{\text{max}}}^2 + \cdots)^{\frac{1}{2}} \tag{2-103}$$

In design practice, modal analysis is employed in conjunction with the response spectra. In this technique, the structure is first decomposed into SDOF systems with respect to the normal coordinates. The maximum response of each SDOF system is derived by reference to the corresponding response spectrum. The base shear obtained from the response is then distributed to all floor levels. Finally, the maximum response of each floor level is computed by the root-sum-square method. In this computation procedure, the mass $\bar{M}_n$ of each SDOF system (defined with respect to the normal coordinates) can be determined from the condition that the sum of base shears of the SDOF system equals the base shear V of the original MDOF system. That is,

$$V = - \sum_{i=1}^{N} m_i(\ddot{v}_i + \ddot{v}_g) = - \sum_{n=1}^{N} \bar{M}_n(\ddot{Y}_{no} + \ddot{v}_g) \tag{2-104}$$

Substituting Eqs. (2-100), (2-101), and (2-86) into the above equation leads to

$$\bar{M}_n = \frac{\left(\sum\limits_{i=1}^{N} m_i \phi_{in}\right)^2}{\sum\limits_{i=1}^{N} m_i \phi_{in}^2} \tag{2-105}$$

$\bar{M}_n$ is called the effective mass, and the sum of all $\bar{M}_n$'s equilibrates the total mass. This implies that the mass $\bar{M}_n$ participates in the nth-mode vibration.

The base shear V_n of the nth mode is distributed to each floor level where the seismic force F_{in} is applied. According to Eq. (2-101), F_{in} is computed as

$$F_{in} = - m_i \phi_{in} \beta_n \omega_n^2 Y_{no} \tag{2-106}$$

and $$V_n = \sum_{i=1}^{N} F_{in} = - \beta_n \omega_n^2 Y_{no} \sum_{i=1}^{N} m_i \phi_{in} \tag{2-107}$$

Then,

$$F_{in} = V_n \frac{m_i \phi_{in}}{\sum\limits_{i=1}^{N} m_i \phi_{in}} \tag{2-108}$$

Use of Eq. (2-96) gives an expression for the displacement at the ith floor level caused by the nth-mode vibration:

$$v_{in} = \phi_{in} \beta_n Y_{no} \tag{2-109}$$

From Eqs. (2-106) and (2-109)

$$v_{in} = - \frac{F_{in}}{\omega_n^2 m_i} \tag{2-110}$$

2.3 Vibration of a One-Dimensional Continuum

2.3.1 Vibration of Shear Beams

2.3.1.1 Free Vibration   Let us consider a slender beam vibrating in the shear mode (Fig. 2-20). The equation of equilibrium of shear forces for a differential segment and the shear-force-versus-deflection relationship are respectively

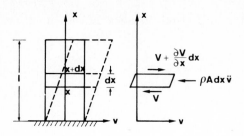

Figure 2-20 Shear vibration of a cantilever beam and forces acting on a differential element.

$$\rho A \, \frac{\partial^2 v}{\partial t^2} = \frac{\partial V}{\partial x} \qquad \text{and} \qquad V = AG \, \frac{\partial v}{\partial x}$$

From these two equations, one can obtain

$$\frac{\partial^2 v}{\partial t^2} = c^2 \frac{\partial^2 v}{\partial x^2}$$

$$c = (G/\rho)^{\frac{1}{2}} \tag{2-111}$$

in which G, ρ, and A are respectively the shear modulus, the mass per unit volume, and the cross-sectional area of the beam. As will be discussed in Sec. 2.3.3, the coefficient c in Eq. (2-111) is the velocity of wave propagation if the equation is regarded as a wave equation.

The number of degrees of freedom of the beam is infinite, which means that the beam possesses an infinite number of natural circular frequencies. Let us suppose that one of those frequencies is ω and that the solution of Eq. (2-111) is expressed as

$$v = \phi(x) \exp (i\omega t) \tag{2-112}$$

By substituting this assumed solution into the governing equation, one can obtain the following form:

$$\frac{d^2\phi}{dx^2} + \frac{\omega^2}{c^2} \phi = 0 \tag{2-113}$$

This form is identical to the equation of free vibration of an SDOF system. The solution of Eq. (2-113) is then

$$\phi = A \cos \frac{\omega}{c} x + B \sin \frac{\omega}{c} x \tag{2-114}$$

Let us assume for the boundary conditions that the beam is clamped at one end ($v = 0$) and free at the other end ($V = 0$). The coefficients A and B in Eq. (2-114) are derived as

$$A = 0 \qquad B \cos (\omega \ell / c) = 0$$

Hence $\omega_n = \dfrac{(2n-1)\pi}{2}\dfrac{c}{\ell}$ $(n = 1, 2, 3, \ldots)$ (2-115)

$$T_n = \frac{2\pi}{\omega_n} = \frac{T_1}{2n-1}$$ (2-116)

The ratios of the second- and third-mode periods to the first-mode period are 1:3 and 1:5, and the modal shapes are as shown in Fig. 2-21.

Equation (2-112) can be rewritten as

$$v_n = (C_n \cos \omega_n t + D_n \sin \omega_n t) \sin \frac{\omega_n}{c} x$$ (2-117)

By substituting all modes of vibration, one can obtain

$$v = \sum_{n=1}^{\infty} (C_n \cos \omega_n t + D_n \sin \omega_n t) \sin \frac{\omega_n}{c} x$$ (2-118)

Coefficients C_n and D_n can be determined from the initial conditions.

2.3.1.2 Forced Vibration   When the beam in Fig. 2-20 is subjected to ground motion, the equation of motion can be expressed as

$$\frac{\partial^2 v}{\partial t^2} - c^2 \frac{\partial^2 v}{\partial x^2} = -\frac{\partial^2 v_g}{\partial t^2}$$ (2-119)

in which v_g is the ground displacement.

The solution of Eq. (2-119) is assumed to be

$$v = \sum_{n=1}^{\infty} \phi_n(x) Y_n(t) \qquad \phi_n = \sin \frac{\omega_n}{c} x$$ (2-120)

where $Y_n(t)$ and $\phi_n(x)$ are a time-varying normal coordinate and a shape function, respectively.

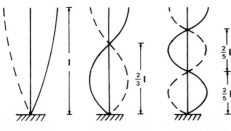

first mode

$T_1 = \dfrac{4\,l}{c}$

second mode

$T_2 = \dfrac{4\,l}{3c}$

third mode

$T_3 = \dfrac{4\,l}{5c}$

Figure 2-21 Modes of shear vibration of a cantilever beam.

Substituting Eq. (2-120) into Eq. (2-119) leads to

$$\sum_{n=1}^{\infty} \phi_n \frac{d^2 Y_n}{dt^2} + \sum_{n=1}^{\infty} \omega_n^2 \phi_n Y_n = -\frac{d^2 v_g}{dt^2} \tag{2-121}$$

Multiplying Eq. (2-121) by ϕ_n and integrating the equation from 0 to ℓ with consideration of the orthogonality of the shape functions gives us

$$\frac{d^2 Y_{no}}{dt^2} + \omega_n^2 Y_{no} = -\frac{d^2 v_g}{dt^2}$$

$$Y_n = \beta_n Y_{no} \tag{2-122}$$

$$\beta_n = \frac{\displaystyle\int_0^{\ell} \phi_n dx}{\displaystyle\int_0^{\ell} \phi_n^2 dx}$$

The term β_n is referred to as the *participation factor*. This procedure is analogous to the one employed in solving the equations of motion of an MDOF system. As seen in Eq. (2-122), the governing equation is uncoupled into independent equations of motion with respect to the normal coordinates. Each uncoupled equation is identical to an equation for the forced vibration of an SDOF system.

If the ground displacement v_g is given, one can obtain Y_{no}. By adding the complementary solution to Y_{no}, the final form of the solution is written as

$$v = \sum_{n=1}^{\infty} \left[\phi_n \beta_n Y_{no} + (C_n \cos \omega_n t + D_n \sin \omega_n t) \sin \frac{\omega_n}{c} x \right] \tag{2-123}$$

where C_n and D_n can be determined from initial conditions.

2.3.2 Vibration of Flexural Beams

By supposing that the beam shown in Fig. 2-22 vibrates in the flexural mode, the distributed force p acting on a differential length dx is given as

$$p = \frac{d^2}{dx^2} \left(EI \frac{d^2 v}{dx^2} \right) \tag{2-124}$$

The equation of motion is obtained by replacing p by the inertial force. For a beam having uniform cross section along the length

$$EI \frac{\partial^4 v}{\partial x^4} + \rho A \frac{\partial^2 v}{\partial t^2} = 0 \tag{2-125}$$

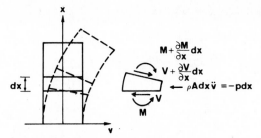

Figure 2-22 Flexural vibration of a cantilever beam and forces acting on a differential element.

in which symbols E, I, ρ, and A are the Young's modulus, moment of inertia, mass per unit volume, and cross-sectional area of the beam, respectively. By taking Eq. (2-112) as a complementary solution of Eq. (2-125), one can obtain

$$\frac{d^4\phi}{dx^4} - \frac{\rho A}{E I}\omega^2\phi = 0 \tag{2-126}$$

The solution is then

$$\phi = A\cos ax + B\sin ax + C\cosh ax + D\sinh ax \tag{2-127}$$

where

$$a = \left(\frac{\rho A\omega^2}{EI}\right)^{\frac{1}{4}} \tag{2-128}$$

As the support conditions at one end of the beam are taken to be clamped and at the other to be free (Fig. 2-22), the following equation must be satisfied with $\phi(0) = \phi'(0) = \phi''(\ell) = \phi'''(\ell) = 0$:

$$\cos a\ell \cosh a\ell + 1 = 0 \tag{2-129}$$

If $a\ell$ is large, $\cosh a\ell$ becomes large enough to assume that $\cos a\ell = 0$. Hence Eq. (2-129) can reasonably be approximated as

$$a\ell \doteq (n - \tfrac{1}{2})\pi$$

and

$$T_n = \frac{2\pi}{\omega_n} = \frac{2\ell^2}{(n - \tfrac{1}{2})^2\pi}\left(\frac{\rho A}{EI}\right)^{\frac{1}{2}} \tag{2-130}$$

The ratios of the second- and third-mode periods to the first-mode period are approximately $(1{:}3)^2$ and $(1{:}5)^2$, respectively, and their modal shapes are as shown in Fig. 2-23. The complementary solution is determined by the use of ω_n [calculated by Eq. (2-130)] and Eqs. (2-127), (2-128), and (2-112):

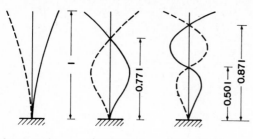

first mode   second mode   third mode

Figure 2-23   Modes of flexural vibration of a cantilever beam.

$$v_n = (C_n \cos \omega_n t + D_n \sin \omega_n t)\phi_n \tag{2-131}$$

$$v = \sum_{n=1}^{\infty} (C_n \cos \omega_n t + D_n \sin \omega_n t)\phi_n \tag{2-132}$$

Coefficients C_n and D_n are determined from the initial conditions.

2.3.3   Wave Propagation in a One-Dimensional Body

As discussed in Sec. 2.3.1, the shear vibration of a beam can be treated as a wave-propagation problem. Wave-propagation theory can be extended to the vibration of a high-rise building structure under ground motion or to the vibration of a slender column. The best application of this theory is to propagation of shear waves through the ground in a vertical direction. By regarding Eq. (2-111) as an equation of wave propagation, one readily finds that the solution is

$$v = f\left(t - \frac{x}{c}\right) + g\left(t + \frac{x}{c}\right) \tag{2-133}$$

As indicated in Fig. 2-24, the first term on the right-hand side of Eq. (2-133) represents a wave propagating forward, whereas the second term represents a wave propagating backward. These two terms are often referred to as the forward- and backward-propagation waves. Symbol c in the equation is the velocity of wave propagation. The relationship between the functions f and g is determined from the boundary conditions. An incident shear wave propagating upward in the ground turns into a reflecting wave when the wave reaches the ground surface (Fig. 2-25). At the surface ($x = 0$), the shear stress should be zero, which means that $G\, \partial v/\partial x = 0$. Substituting this condition into Eq. (2-133), we obtain

$$\frac{\partial f}{\partial t} = \frac{\partial g}{\partial t}$$

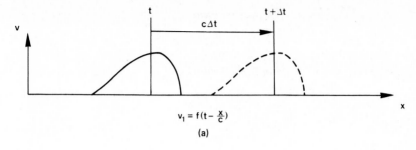

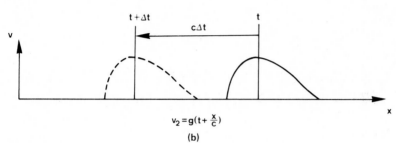

Figure 2-24   Wave propagation. *(a)* Forward-propagating wave. *(b)* Backward-propagating wave.

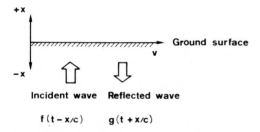

Figure 2-25   Reflection of a wave at the ground surface.

This equation means that $f = g$, and so

$$v = f\left(t - \frac{x}{c}\right) + f\left(t + \frac{x}{c}\right) \tag{2-134}$$

By assuming that the waveform at the surface is v_g,

$$v_g = 2f(t) \tag{2-135}$$

This expression implies that an observed waveform is twice as large as the input incident wave. If the waveform at the ground surface v_g is known,

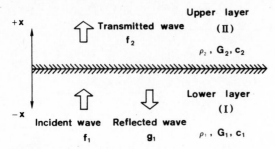

Figure 2-26 Transmission and reflection of a wave at the interface between two layers.

the wave at an arbitrary point in the ground can be obtained from Eqs. (2-134) and (2-135) as

$$v(t,x) = \frac{1}{2}\left[v_g\left(t - \frac{x}{c}\right) + v_g\left(t + \frac{x}{c}\right)\right] \tag{2-136}$$

This equation indicates that the displacement at time t is the average of the displacements of the surface at times earlier and later by t.

When the ground is composed of two layers as shown in Fig. 2-26, part of the wave propagating upward in the lower layer passes through the boundary to the upper layer, while the rest is reflected at the boundary. By supposing that the densities are ρ_1 and ρ_2, the shear moduli are G_1 and G_2, and the velocities of wave propagation are c_1 and c_2 for the lower and upper layers, respectively, then for the waves propagating through the two layers we have expressed as

$$v_1 = f_1\left(t - \frac{x}{c_1}\right) + g_1\left(t + \frac{x}{c_1}\right) \tag{2-137}$$
$$c_1 = (G_1/\rho_1)^{\frac{1}{2}}$$

$$v_2 = f_2\left(t - \frac{x}{c_2}\right) \tag{2-138}$$
$$c_2 = (G_2/\rho_2)^{\frac{1}{2}}$$

At the boundary surface, displacements and shear stresses for the two layers are identical. The compatibility conditions are then

$$v_1 \big|_{x=0} = v_2 \big|_{x=0}$$
$$G_1 \frac{\partial v_1}{\partial x}\bigg|_{x=0} = G_2 \frac{\partial v_2}{\partial x}\bigg|_{x=0} \tag{2-139}$$

By substituting Eqs. (2-137) and (2-138) into Eq. (2-139) and further considering the second equation of Eq. (2-111), we obtain

$$f_2\left(t - \frac{x}{c_2}\right) = \frac{2}{1 + \alpha} f_1\left(t - \frac{x}{c_1}\right)$$

$$g_1\left(t + \frac{x}{c_1}\right) = \frac{1 - \alpha}{1 + \alpha} f_1\left(t + \frac{x}{c_1}\right) \tag{2-140}$$

$$\alpha = \frac{\rho_2 c_2}{\rho_1 c_1} \qquad \beta = \frac{1 - \alpha}{1 + \alpha} \qquad \gamma = \frac{2}{1 + \alpha}$$

In these equations, α, β, and γ are the wave-propagation impedance, the reflection coefficient, and the transmission coefficient, respectively. The incident wave, transmitted wave, and reflected wave have an identical waveform. When $\alpha > 1$ and $\beta < 0$, it is found that the reflected wave is skew-phased relative to the incident wave (Fig. 2-27).

Let us consider the problem illustrated in Fig. 2-28, in which the ground comprises bedrock (I) and a surface layer (II) and the relationship between the incident wave in layer I and the actual displacement of the ground surface is to be examined. According to Eq. (2-136), the waveform

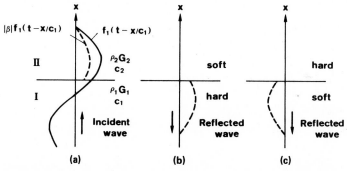

Figure 2-27 Reflection of a wave. (*a*) Incident-wave propagation. (*b*) $\alpha < 1$. (*c*) $\alpha > 1$.

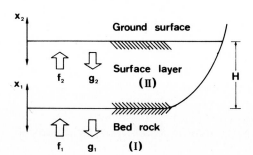

Figure 2-28 Wave propagation in a single layer on bedrock.

at point x_2 in the surface layer can be expressed in terms of the waveform at the ground v_g as

$$v_2(t, x_2) = \frac{1}{2}\left[v_g\left(t - \frac{x_2}{c_2}\right) + v_g\left(t + \frac{x_2}{c_2}\right)\right] \tag{2-141}$$

The waveform at point x_1 in the bedrock is given as

$$v_1(t, x_1) = f_1\left(t - \frac{x_1}{c_1}\right) + g_1\left(t + \frac{x_1}{c_1}\right) \tag{2-142}$$

The compatibility conditions at the boundary between the two layers are

$$v_2(t, -H) = v_1(t, 0)$$

$$G_2\frac{\partial}{\partial x_2}v_2(t, -H) = G_1\frac{\partial}{\partial x_1}v_1(t, 0) \tag{2-143}$$

According to Eqs. (2-141) through (2-143),

$$f_1(t) = \frac{1}{4}\left[\left(1 + \alpha\right)v_g\left(t + \frac{H}{c_2}\right) + (1 - \alpha)\,v_g\left(t - \frac{H}{c_2}\right)\right] \tag{2-144}$$

By use of this equation, the form of the incident wave transmitted through the boundary to the surface layer can be derived if the waveform at the ground surface is known (Tajimi, 1965). By assuming that the waveform at the surface v_g is $A_g \exp(i\omega t)$ and the incident wave $f_1(t)$ is $a \exp(i\omega t)$, one can obtain a as

$$a = \frac{A_g}{4}[(1 + \alpha)\exp(i\omega H/c_2) + (1 - \alpha)\exp(-i\omega H/c_2)] \tag{2-145}$$

$$= \frac{A_g}{2}\left(\cos\frac{\omega H}{c_2} + i\alpha\sin\frac{\omega H}{c_2}\right)$$

The ratio of the amplitude A_g at the ground surface to the amplitude $2a$ at the boundary, provided there is no surface layer, can be written as

$$\left|\frac{A_g}{2a}\right| = \left(\cos^2\frac{\omega H}{c_2} + \alpha^2\sin^2\frac{\omega H}{c_2}\right)^{-\frac{1}{2}} \tag{2-146}$$

This ratio expresses the change in amplitude of the incident wave caused by the surface layer. Figure 2-29 shows the relationship between ω and the amplification with the impedance ratio α as a parameter. When the circular frequency ω of an incident wave coincides with one of the natural

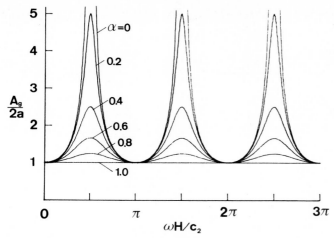

Figure 2-29   Amplification characteristics of the surface layer.

circular frequencies of the surface layer, i.e., $\pi c_2/2H$, $3\pi c_2/2H$, $5\pi c_2/2H$, etc., a resonance condition occurs:

$$\omega_n = \frac{(2n-1)\pi c_2}{2H} \qquad (n = 1, 2, 3, \ldots) \tag{2-147}$$

in which ω_n is the nth natural circular frequency of the surface layer. The frequency can be converted to a natural period as

$$T_n = \frac{1}{2n-1} \frac{4H}{c_2} \tag{2-148}$$

This period T_n is referred to as the *predominant period*. The discussion so far has focused on the case of a single homogeneous boundary layer. The technique to obtain the predominant period discussed above can readily be extended to a case in which several surface layers exist on bedrock (Cherry, 1974). Figure 2-30 presents amplification characteristics of incident waves when the ground is complex in configuration and has viscous damping, showing significant change in response due to the ground properties (Toki, 1981). When an incident wave is not sinusoidal but random (like an earthquake wave), a technique involving the Fourier transform is useful. First, the random wave is decomposed into a complex trigonometric series. For each term, the technique employed to derive the response is that discussed above. Finally, the response is obtained by means of the inverse Fourier transform. Detailed discussion of this technique is given in Cherry (1974).

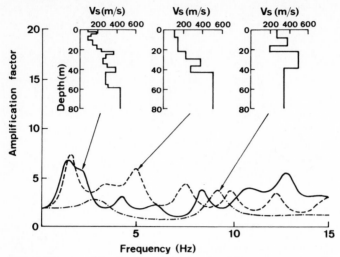

Figure 2-30 Amplification characteristics of multilayers. [*From K. Toki, Earthquake Resistant Analysis of Structures, Gihodo Shuppan Co., Tokyo, 1981 (in Japanese).*]

2.4 Rocking Vibration and Torsional Vibration

2.4.1 Modeling of Soil

Since soil is not perfectly rigid, the ground near a building deforms in response to the building vibration; that is, the building and the ground interact with each other under earthquake disturbances. Often a building not only vibrates in a horizontal direction but also is subjected to rotational vibration, which is usually called *rocking vibration*. To study the translational and rotational vibrations in building structures, we must first introduce suitable models of the ground. Many models have been proposed. Some are relatively simple, whereas others need rigorous formulation.

1. *Spring models of ground.* Probably the simplest model for analysis of the rocking motion of a building during ground disturbances is the spring. In this model, the building is assumed to be supported by springs that represent the characteristics of the ground as shown in Fig. 2-31a. The spring which resists the rotation of the building is identified as the rocking spring. A dashpot can be included if some viscous damping is expected in the ground. The spring constant can be estimated in two ways, experimentally and theoretically. For experimental estimation, the ground is excited by a vibration generator.

In the theoretical approach, it is assumed that the ground is a semi-infinite body and that a dynamic force is applied to the foundation. The

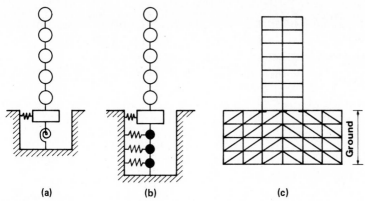

Figure 2-31 Modeling of soil and structures. *(a)* Mass model (single-mass representation of soil). *(b)* Mass model (multimass representation of soil). *(c)* Finite-element model.

equivalent stiffness and viscous damping then can be computed on the basis of the phase difference between the force and the resultant deformation. An example of spring constants and damping coefficients computed by this method is given in Newmark and Rosenblueth (1971). In such a purely theoretical approach, the ground is assumed to be a semi-infinite elastic body composed of at most several soil layers. Applicability of the results is therefore rather limited.

2. *Lumped-mass model of ground.* In this model, the ground is represented by vertically linked lumped masses as shown in Fig. 2-31*b*. Each lumped mass, with its spring constant and damping coefficient, represents one layer of the ground. These properties are difficult to determine, however, and the model does not take energy dissipation into account. Furthermore, the assumption that the surrounding soil is perfectly rigid is often very questionable.

3. *Semi-infinite-body model.* The ground is assumed to be a uniform elastic or viscoelastic semi-infinite body. Radiation damping can be included, and the damping effect of the soil can also be incorporated into the analysis by assuming that the ground is viscoelastic.

4. *Finite-element model.* The ground is discretized into finite elements (Fig. 2-31*c*). The nonuniformity of soil properties is allowed for by assigning different material properties to each finite element. Inelastic soil behavior can be considered by means of nonlinear finite-element computation. One disadvantage of this model is the high cost of analysis.

Since finite-element analysis is costly, the discretization should be carefully selected. If the ground consists of layers each of which spreads in the horizontal plane with uniform material properties, one-dimensional discretization is suitable. If the soil is confined in a long, narrow valley, a two-dimensional model is useful. If the stratum or the contact area between the building and the ground is symmetical about the vertical axis of revolution, axisymmetrical analysis is useful. In any case, a rigid boundary which confines the energy dissipation of the ground must be defined in the discretization. Moreover, in model 2, the assumed input ground motion is not identical with the real condition.

In sections to follow, the simplest spring model is explained in greater detail. Section 2.5.2 depicts the interaction between ground and buildings, particularly that associated with radiation damping.

2.4.2 Periods and Modes of Rocking Vibration

As shown in Fig. 2-32, the motion of a rigid body supported by springs can be expressed in terms of the translation v of the centroid of the body and the rotation θ about the centroid.

From force equilibrium in the horizontal direction (y direction in Fig. 2-32) and moment equilibrium about the centroid, we obtain

$$m \, \ddot{v} = - k_h(v - s\theta)$$
$$I_G \ddot{\theta} = - k_r\theta + k_h(v - s\theta)s$$

where I_G indicates the polar moment of inertia about the centroid of the body and k_h and k_r respectively indicate spring coefficients of the horizontal and rotational springs.

The above equations can be rewritten as

$$m\ddot{v} + k_h v - k_h s\theta = 0$$
$$I_G \ddot{\theta} - k_h sv + (k_r + k_h s^2)\theta = 0$$

$$(2\text{-}149)$$

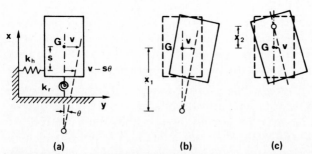

Figure 2-32 Rocking vibration of a rigid body. (a) Rigid-body representation. (b) First-mode vibration. (c) Second-mode vibration.

By assuming that the unknown horizontal displacement v and rotation θ are $\hat{v}$ exp $(i\omega t)$ and θ *exp (iωt)* and substituting them into Eq. (2-149), we derive the following equations:

$$(-\omega^2 m + k_h)\hat{v} - k_h s\hat{\theta} = 0$$
$$-k_h s\hat{v} + (-\omega^2 I_G + k_r + k_h s^2)\hat{\theta} = 0 \tag{2-150}$$

To obtain nontrivial solutions for $\hat{v}$ and $\hat{\theta}$ the determinant of the coefficient matrix of the equations should be zero. Thus,

$$\omega^4 - \left(1 + \frac{e_0^2}{i_0^2} + \frac{s^2}{i_0^2}\right)\omega_h^2\omega^2 + \frac{e_0^2}{i_0^2}\omega_h^4 = 0 \tag{2-151}$$

in which $\omega_h^2 = k_h/m$, $i_0^2 = I_G/m$, and $e_0^2 = k_r/k_h$. From this equation, the natural circular frequencies ω_1 and ω_2 are determined as

$$\left.\begin{array}{l}\omega_1^2/\omega_h^2\\\omega_2^2/\omega_h^2\end{array}\right\} = \frac{1}{2}\left\{1 + \frac{e_0^2}{i_0^2} + \frac{s^2}{i_0^2} \mp \left[\left(1 + \frac{e_0^2}{i_0^2} + \frac{s^2}{i_0^2}\right)^2 - 4\frac{e_0^2}{i_0^2}\right]^{\frac{1}{2}}\right\} \tag{2-152}$$

By substituting these frequencies back into Eq. (2-150), we obtain rocking-mode shapes for the rigid body. The rotation centers of rocking for the first and second rocking modes can be expressed as

$$x_1 = \frac{s}{1 - (\omega_1/\omega_h)^2}$$
$$x_2 = \frac{s}{1 - (\omega_2/\omega_h)^2} \tag{2-153}$$

As illustrated in Fig. 2-32, the center of rotation of the first rocking mode lies below the centroid of the body, whereas that of the second mode is above the centroid (Tajimi, 1965).

An example of a one-mass system in which the base can translate and rotate is shown in Fig. 2-33. Here, the base is represented as a mass m_0, supported by a translational spring k_h, and a rotational spring k_r. The

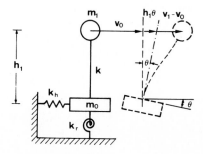

Figure 2-33 Rocking vibration of a two-mass system.

system has two masses, m_1 and m_0. The equations of motion of this system are written as

$$m_0\ddot{v}_0 + (k_h + k)v_0 - kv_1 = 0$$
$$m_1(\ddot{v}_1 + h_1\ddot{\theta}) - kv_0 + kv_1 = 0 \qquad (2\text{-}154)$$
$$I_G\ddot{\theta} + m_1(\ddot{v}_1 + h_1\ddot{\theta})h_1 + k_r\theta = 0$$

or in matrix form

$$\mathbf{m}\ddot{\mathbf{v}} + \mathbf{k}\mathbf{v} = \mathbf{0} \qquad (2\text{-}155)$$

where
$$\mathbf{m} = \begin{bmatrix} m_0 & 0 & 0 \\ 0 & m_1 & m_1 h_1 \\ 0 & m_1 h_1 & I_G + m_1 h_1^2 \end{bmatrix} \qquad \mathbf{v} = \begin{Bmatrix} v_0 \\ v_1 \\ \theta \end{Bmatrix}$$

$$\mathbf{k} = \begin{bmatrix} k_h + k & -k & 0 \\ -k & k & 0 \\ 0 & 0 & k_r \end{bmatrix} \qquad (2\text{-}156)$$

As Eq. (2-155) is merely the equation of motion of a typical MDOF system, the solution procedure is identical with that explained in Sec. 2.2.4.

2.4.3 Rocking Vibration under Ground Motions

Let us come back to Fig. 2-32. By supposing that the base is subjected to motion $\ddot{v}_g$ (in terms of acceleration), the equations of motion are the same as Eq. (2-149) except that $-m\ddot{v}_g$ appears on the right-hand side of the first equation instead of zero. The equations are expressed in matrix form as

$$\mathbf{m}\ddot{\mathbf{v}} + \mathbf{k}\mathbf{v} = -\mathbf{m}\mathbf{f}\ddot{v}_g \qquad (2\text{-}157)$$

$$\mathbf{f} = \begin{Bmatrix} 1 \\ 0 \end{Bmatrix} \qquad (2\text{-}158)$$

These equations are solved in the same manner as the equations of motion of an undamped MDOF system under forced vibration [Eq. (2-60)]. When a base motion $\ddot{v}_g$ is applied to the system shown in Fig. 2-33, the equations of motion turn out to be identical with Eq. (2-157). In this case, Eq. (2-158) should be modified to

$$\mathbf{f} = \begin{Bmatrix} 1 \\ 1 \\ 0 \end{Bmatrix} \qquad (2\text{-}159)$$

2.4.4 Periods and Modes of Torsional Vibration

If a space structure is subjected to horizontal ground motion and the center of mass of the structure does not coincide with the center of reaction of the structure, torsional vibration occurs. Let us consider a one-story space frame with a rigid diaphragm as illustrated in Fig. 2-34. The displacement of an arbitrary point of the space frame is expressed in terms of the horizontal displacements u and v of the centroid of the frame and the rotation θ about the centroid. The displacements of the center of stiffness C (identical with the reaction center) in Fig. 2-34 are $u - e_y\theta$ and $v + e_x\theta$ in the x and y directions. If the story stiffnesses are k_x and k_y in the x and y directions, then the equations of motion at the center of stiffness are

$$-m\ddot{u} = k_x(u - e_y\theta)$$
$$-m\ddot{v} = k_y(v + e_x\theta)$$

The inertial moment is $-I_G\ddot{\theta}$, where I_G is the polar moment of inertia about the centroid. This moment then is in equilibrium with a combination of the torsional resistance $k_\theta\theta$ about the center of stiffness and the translational resistances $-k_x(u - e_y\theta)e_y$ and $k_y(v + e_x\theta)e_x$ at the center of stiffness. Thus,

$$-I_G\ddot{\theta} = k_\theta\theta - k_x(u - e_y\theta)e_y + k_y(v + e_x\theta)e_x$$

Here k_θ is the torsional stiffness of the frame about the center of stiffness. The equations of free vibration are then written as

$$m\ddot{u} + k_x(u - e_y\theta) = 0$$
$$m\ddot{v} + k_y(v + e_x\theta) = 0 \tag{2-160}$$
$$I_G\ddot{\theta} + (k_\theta + k_xe_y^2 + k_ye_x^2)\theta - k_xe_yu + k_ye_xv = 0$$

By assuming that the solutions are $u = \hat{u}\exp(i\omega t)$, $v = \hat{v}\exp(i\omega t)$, $\theta = \hat{\theta}\exp(i\omega t)$ and substituting these into the above equations, we obtain three homogeneous equations with $\hat{u}$, $\hat{v}$, and $\hat{\theta}$ as unknowns. From the basic eigensolution procedure, three natural circular frequencies can be derived. These frequencies correspond to the two translational-vibration

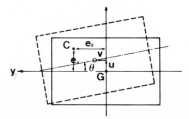

Figure 2-34 Torsional vibration.

modes in the principal axes and the torsional-vibration mode. If the center of stiffness C is on the y axis, the vibration in the y direction is uncoupled from the other two. The equations of motion then are simplified to

$$m\ddot{u} + k_x u - k_x e_y \theta = 0$$
$$I_G \ddot{\theta} + (k_\theta + k_x e_y^2)\theta - k_x e_y u = 0 \tag{2-161}$$

The above equations have identical forms with those that govern the rocking motion [described in Eq. (2-149)]. Analogously to the rocking motion, the natural circular frequencies of torsional motion given in Eq. (2-161) are found as (Tajimi, 1965)

$$\left.\begin{array}{c}\omega_1^2/\omega_x^2 \\ \omega_2^2/\omega_x^2\end{array}\right\} = \frac{1}{2}\left\{ 1 + \frac{e_0^2}{i_0^2} + \frac{e_y^2}{i_0^2} \mp \left[\left(1 + \frac{e_0^2}{i_0^2} + \frac{e_y^2}{i_0^2}\right)^2 - 4\frac{e_0^2}{i_0^2}\right]^{\frac{1}{2}} \right\} \tag{2-162}$$

where $\quad i_0^2 = \dfrac{I_G}{m}\quad$ and $\quad e_0^2 = \dfrac{k_\theta}{k_x}\quad$ and $\quad \omega_x^2 = \dfrac{k_x}{m}$ $\qquad$ (2-163)

When e_y is zero in Eq. (2-162), $\omega_1 = \omega_x e_0/i_0$, and $\omega_2 = \omega_x$. e_0 in Eq. (2-163) is the radius of gyration of the horizontal force resisting components about the center of stiffness. If $e_0 > i_0$, as when walls are arranged along the periphery of a frame, $\omega_2 < \omega_1$. This means that the translational-vibration mode is the first mode. On the other hand, if $e_0 < i_0$, as when a stiff core is arranged in the middle of a frame, the torsional-vibration mode is the first mode. As indicated in this example, even a one-story frame has as many as three degrees of freedom. The analysis of torsional motion is very complicated for a multistory frame.

2.4.5 Torsional Vibration of Space Structures

When the one-story frame in Fig. 2-34 is subjected to time-varying ground acceleration in the direction $\ddot{u}_g$, the equations of motion are the same as Eq. (2-160) except that the right-hand side of the first equation (zero) is replaced by $-m\ddot{u}_g$. The equations in matrix form are

$$\mathbf{m\ddot{u} + ku} = -\mathbf{m f}\ddot{u}_g \tag{2-164}$$

where $\qquad \mathbf{f} = \left\{\begin{array}{c} 1 \\ 0 \\ 0 \end{array}\right\}$

The equations are essentially the same as those of an undamped MDOF system subjected to forced vibration. According to Shiga's study (1976) on the torsional response of a one-story frame subjected to realistic earthquake ground motion, dynamic eccentricity is about twice as large as static eccentricity when e_0/i_0, defined in Eq. (2-163), is 1.0 to 1.5. If this value is 2.0, the ratio of dynamic to static eccentricity is reduced to about 1.5 (Shiga, 1976). As seen from this study, the effect of torsional vibration on overall response depends largely on torsional stiffness. More detailed descriptions are given in Müller and Keintzel (1978).

2.5 Dynamic Characteristics of Structures

2.5.1 Restoring Force

To study the inelastic response of a discrete mass system, we must set up a mathematical model of restoring-force characteristics and hence define the relationship between the story shear force and the story deflection.

For a progressive sequence of loadings and unloadings, the line join-ing the peak points in the load-deflection curve of each loading sequence is termed the *skeleton curve*. In many cases the skeleton curve coincides with the monotonic loading curve. The curve obtained under force reversals is called the *hysteresis curve*. The hysteresis curve is significantly affected by materials and structural type (this will be discussed in detail in Chap. 3). Most mathematical models are simplified according to the required level of analysis.

Figure 2-35a shows the most popular bilinear hysteresis model. When line *AB* in the figure has a positive slope, the model is *positive bilinear;* it is *negative bilinear* if line *AB* has a negative slope.

If the slope is zero, the model is identical with the elastoplastic model. An elastoplastic or positive bilinear model is often used to represent the restoring-force characteristics of a steel frame. When the frame is sub-jected to high axial force, a negative bilinear model is sometimes useful. For simplicity, this bilinear model is sometimes used for a reinforced-concrete frame. The displacement response of a mass model under earth-quake disturbances is affected greatly by the selection of parameters in the model. The displacement response is reduced with increasing slope in the second branch (line *AB*) if the yield level is unchanged. As the slope decreases, the response increases and tends to be biased in one direction if the slope is negative.

Figure 2-35b shows the trilinear model. Lines *OABC* constitute the skeleton curve. Line *CD* is parallel to and twice as long as line *OA*. Analogously, line *DE* is parallel to and twice as long as line *AB*. This model is usually used for reinforced-concrete and composite steel and reinforced-concrete frames. In this case, points *A* and *B* correspond to the points of cracking and yielding.

Figure 2.35c is a hysteresis model having the Ramberg-Osgood curve

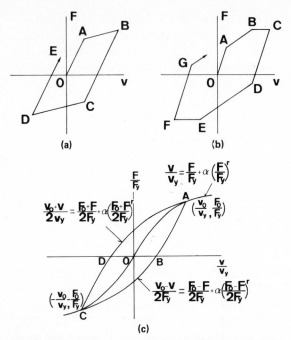

Figure 2-35   Masing-type models. *(a)* Bilinear model. *(b)* Trilinear model. *(c)* Ramberg-Osgood model.

as the skeleton curve. It was first used by Jennings (1964) for his dynamic-response analysis. Curve *ABC* in this figure is obtained by inverting line *OA* and extending it so that its abscissa and ordinate are twice those of *OA*. Models *a, b,* and *c,* which are all essentially the same in the relation between the skeleton curve and the hysteresis curve, are classified as the Masing type. Model *c,* the most realistic of the three, can represent the Bauschinger effect and the effect of sequential yielding of members. The complexity of the hysteresis law is a disadvantage of this model, however, and models *a* and *b* are more frequently used for design purposes.

Figure 2-36 is another model, called the *degrading type.* This model allows for the effect of stiffness degradation caused by load reversals in inelastic ranges in a reinforced-concrete frame that yields by flexure. Many models of this type have been proposed (Clough and Johnston, 1966; Takeda, Sozen, and Nielsen, 1970). As an example, in the degrading bilinear model (Fig. 2-36*a*) proposed by Clough and Johnston line *BC* is parallel to line *OA*. From point *C* the stiffness changes, heading directly toward the yield point *D* (in the negative direction). Line *EF* is parallel to line *OD* but changes the slope at point *F*. The line then proceeds to point *B*, the load-reversal point in the previous cycle. The displacement re-

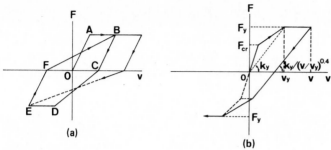

(a)

(b)

Figure 2-36 Degrading models. *(a)* Degrading bilinear model by Clough and Johnston. [*Clough and Johnston, Effect of stiffness degradation on earthquake ductility requirements*, Proc. Second Japan Earthquake Eng. Symp., *Tokyo, 227–232 (1966).*] *(b)* Degrading trilinear model by Takeda. [*T. Takeda, M. A. Sozen, and N. N. Nielsen, Reinforced concrete response to simulated earthquakes*, J. Struct. Div., Am. Soc. Civ. Eng., *96(ST-12), 2557–2573 (1970).*]

sponse increases as stiffness is degraded or as the stiffness of the second branch of the skeleton curve becomes smaller. The hysteresis model in Fig. 2-37*a* (Tanabashi and Kaneta, 1962; Iwan, 1965) is called the *slip-type model*. By combining this model with the bilinear model, the hysteresis curve shown in Fig. 2-37*b* can be obtained. This combined model is very useful for a bolt connection in a steel structure, in which bolts are expected to slip. This model is also often used to represent a bracing member that has a significant buckling effect and also a reinforced-concrete member in which shear distortion dominates overall behavior. Other hysteresis models such as that including capacity degradation due

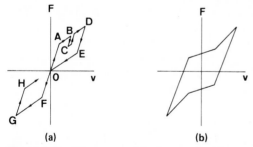

(a)

(b)

Figure 2-37 Slip-type models. *(a)* Double bilinear model by Tanabashi and Kaneta [*R. Tanabashi and K. Kaneta, On the relation between the restoring force characteristics of structures and the pattern of earthquake ground motions*, Proc. First Japan Earthquake Eng. Symp., *Tokyo, 57–62 (1962)*] and Iwan [*W. D. Iwan, The steady state response of the double bilinear hysteretic model*, Trans. Am. Soc. Mech. Eng., J. Appl. Mech., *32, 921–925 (1965)*]. *(b)* Slip-type model.

to severe shear reversals in reinforced-concrete members have been proposed. In general, however, it is difficult to model such complicated behavior.

2.5.2 Damping Characteristics

2.5.2.1 Types of Damping The damping of structures under earthquake disturbances consists of external viscous damping, internal viscous damping, body-friction damping, hysteresis damping, and damping by radiation to the ground.

External Viscous Damping. This is the damping caused by the air or water surrounding a structure. It is negligibly small in comparison with other types of damping.

Internal Viscous Damping. This is the damping associated with material viscosity. It is proportional to velocity, so that the damping ratio increases in proportion to the natural frequency of the structure. Internal viscous damping is readily included in dynamic analysis by introducing a dashpot as described in Sec. 2.1. It is frequently used to represent all kinds of damping.

Body-Friction Damping. This damping which is also called *Coulomb damping,* occurs because of friction at connections or support points. It is constant regardless of the velocity or amount of displacement, and it is usually treated either as internal viscous damping when the level of displacement is small or as hysteresis damping when it is large. Body friction is large in infilled masonry walls when the walls crack and provides very effective seismic resistance.

Hysteresis Damping. This damping takes place when a structure is subjected to load reversals in the inelastic range. As shown in Fig. 2-38, a one-cycle hysteresis loop swells outward. Energy corresponding to the area of the loop is dissipated in the cycle. This dissipation in energy is

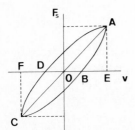

Figure 2-38 Load-deflection hysteresis loop.

defined as *hysteretic damping*. It is unaffected by the velocity of the structure but increases with the level of displacement. In Fig. 2-38 a one-cycle hysteresis loop is shown in terms of the force-deflection relationship. The energy input from point D to point A is given by the area defined by points DAE. When the structure moves from point A to point B, the energy represented by area BAE is emitted. The same is true between points B and C and between points C and D. As a result, the energy corresponding to the area $ABCD$ is dissipated in one cycle of load reversal. Such damping, associated with the hysteresis loop in the inelastic range, can be incorporated into analysis by assuming a spring that reflects nonelastic restoring-force characteristics as indicated in Sec. 2.6. Analysis with such a spring model usually is very complicated. Instead, hysteresis damping is often replaced by *equivalent viscous damping*, and an elastic analysis is performed. This approach was first proposed by Jacobsen (1930). The concept is that an inelastic spring system vibrating under stationary sinusoidal base motion is replaced by an elastic damped spring system which is subjected to the same motion and has the same natural frequency and energy-dissipation capacity as the inelastic system. Let us consider a system having restoring-force characteristics as shown in Fig. 2-38. In this system, the spring constant varies with the force. In the equivalent damped mass system, the spring constant is assumed to be the one represented by the line AOC, and the equivalent viscous damping is given by

$$\xi_{eq} \equiv \frac{1}{2\pi} \frac{\text{area of loop } ABCDA}{\Delta OAE + \Delta OCF} \equiv \frac{1}{2\pi} \frac{\Delta W}{W} \tag{2-165}$$

This equation is derived by equating the area $ABCDA$ with the energy dissipation by viscous damping ξ_{eq}. Various proposals have been made for determining the area in the denominator of Eq. (2-165) when the skeleton curve is nonlinear (Rea, Clough, et al., 1969). Care must be taken since the value of ξ_{eq} differs according to each proposal.

Hysteresis damping caused by plastic deformation is frequently very large. As an example, for a system which has elastic, perfectly plastic hysteresis characteristics, ξ_{eq} is calculated as 0.16 and 0.21 if the maximum deflection is 1.5 and 2.0 times as large as the yield deflection. Use of this equivalent damped system is reasonable as long as ξ_{eq} is small, but the error grows as ξ_{eq} increases (Jennings, 1968).

Radiation Damping. When a building structure vibrates, elastic waves propagate through the semi-infinitely extended ground on which the structure is built. The energy inputted into the structure is dissipated by this wave propagation. The dissipation in energy, defined as *radiation damping*, is a function of the elastic constant E, the density ρ, the Poisson ratio ν of the ground, the mass of the structure per unit area m/A, and the ratio of the spring constant of the structure to the mass k/m. Radiation damping increases and eventually the structural response decreases as the structure becomes stiffer, the ground more flexible, and embedment deeper. Radiation damping is smaller for higher-mode vibration, which is the reverse of the case of internal viscous damping (Katayama, 1969). As Veletsos and Nair (1975) mention, radiation and ground hysteresis damping are not additive to structural damping.

Hysteresis Damping around the Foundation. This is part of external damping and is caused by inelastic deformation of the ground in the vicinity of the foundation.

2.5.2.2 Damping Values for Building Structures   In most dynamic-response analyses of building structures, all the various sources of damping are represented by viscous damping. In this case, hysteresis damping is taken into account by introducing an equivalent viscous damping. This simplification, however, leads to erroneous results when the level of deflection is large, as discussed in the preceding section. In more refined analyses, hysteresis damping is often considered in stiffness representation by the use of inelastic restoring-force characteristics. When high-rise building structures are analyzed for their earthquake response, damping-ratio values of 0.02 and of 0.03 to 0.05 are used for steel and reinforced-concrete or composite steel and reinforced-concrete structures in Japanese practice. Damping ratios corresponding to higher modes are assumed to increase in proportion to natural frequencies. Dowrick (1977) provides a list of damping values used for various types of building structures. Figure 2-39 shows damping ratios obtained from vibration tests of existing buildings (Aoyama, 1980). It is clear from this figure that values scatter over a wide range, and it is difficult to make firm suggestions.

2.5.3 Calculation of Dynamic Characteristics of Model Structures

In the dynamic-response analysis of an MDOF system, the natural frequencies and mode shapes of the system must be determined. The method described in Sec. 2.2.2 can be used to compute frequencies and modal shapes, but it is very time-consuming and not practical if degrees of freedom exceed four. Various approximate techniques have been devised to reduce the complexity of the problem. Nowadays, many computer programs are available for calculating the eigenvalue of large matrices.

Two of the most popular approximate techniques are the Stodola method and the Holzer method. With either method, eigencalculations for

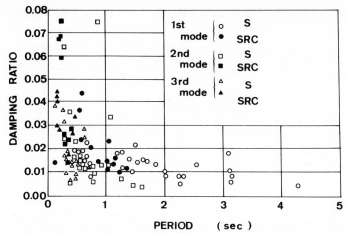

Figure 2-39 Damping ratios measured in existing buildings. [*By the Architectural Institute of Japan, from H. Aoyama, Trends in the earthquake resistant design of SRC highrise buildings, in B. Kato and L.-W. Lu (eds.),* Developments in Composite and Mixed Construction, *Proc. U.S.A.— Japan Seminar on Composite Structures and Mixed Construction Systems, Gihodo Shuppan Co., Tokyo, 1980, pp. 171–181.*]

systems with up to 10 degrees of freedom can readily be made. In the Stodola method, modal shapes are assumed and updated successively until the assumed shapes are close enough to the exact shapes. The natural frequencies are then computed. On the other hand, in the Holzer method the natural frequencies are first assumed and then updated successively until the true frequencies are obtained. The modal shapes can then be derived from the frequencies. In this method, the natural frequency of an arbitrary mode can be derived first. Readers may wish to refer to Clough and Penzien (1975) for details of these methods.

In some cases, e.g., when the design earthquake force for a structure is being estimated, only a rough value of the fundamental natural period may be needed. One such approximate procedure is the weight method. In this method, the natural period is estimated for the first modal shape, which is taken to be the deflected shape when gravity forces are applied to the structure.

Let us consider again the shear-type continuum and find its fundamental natural period T_1. From Eqs. (2-115) and (2-116),

$$T_1 = 4\left(\frac{\rho\ell^2}{G}\right)^{\frac{1}{2}} \tag{2-166}$$

The deflection δ when gravitational force is applied to the structure in the horizontal direction is given as

$$\delta = \frac{g\rho\ell^2}{2G} \tag{2-167}$$

According to Eqs. (2-166) and (2-167), a term $\rho\ell^2/G$ can be canceled out, and

$$T_1 = \frac{\delta^{\frac{1}{2}}}{5.53}$$

For a frame-type structure, the value in the denominator of this equation is 5, 5.4, and 5.7, respectively, for one-, two-, and three-story buildings. For practical purposes, the following is adequate:

$$T_1 = \frac{\delta^{\frac{1}{2}}}{5.5} \tag{2-168}$$

where δ is expressed in centimeters.

Analogously, the fundamental natural period of a bending-type continuum is given by replacing 5.5 in Eq. (2-168) with 6.2. The relationship between the fundamental natural period and higher-mode natural periods can be expressed by Eqs. (2-116) and (2-130) for shear and bending behavior, respectively. Furthermore, we can find many practical equations that estimate the fundamental natural frequency of buildings (San Francisco ASCE-SA, 1952; Taniguchi, 1960; Otsuki, 1960; Nakagawa, 1960; Kanai, 1962; Housner and Brady, 1963). In these references, natural frequency is expressed as a function of the number of stories, story height, depth of the building, and other factors. Table 2-1 lists some equations prescribed in building-design codes for determining equivalent earthquake force in the aseismic design of building structures.

2.5.4 Dynamic Testing of Structures

In determining the dynamic characteristics of building structures, it is often necessary to conduct experimental tests in addition to theoretical

TABLE 2-1   Fundamental Period of Building, s, as Prescribed in Various Building Codes

	UBC	ATC-3		Japanese code	
		Steel	Concrete	Steel	Concrete
Moment-resisting frame	$0.10N$	$0.035H$	$0.025H$	$0.03\bar{H}$	$0.02\bar{H}$
Other frame	$\dfrac{0.05H}{D^{1/2}}$	$\dfrac{0.05H}{D^{1/2}}$			

NOTE: N = number of stories; H, $\bar{H}$ = height from the base to the highest level of the building, ft and m, respectively; D = overall length, ft, of the building in the direction under consideration.

studies. The test structure should be loaded to failure if ultimate capacity is to be found. Small-scale-model tests are popular for this type of testing.

When the dynamic behavior of a real structure is studied, the motion induced by excitation generators is not sufficient in most cases to cause inelastic structural action. The test is elastic and therefore provides information on the natural period, mode shape, and viscous damping of the structure. Hudson (1970) has a comprehensive review of vibration testing of building structures.

2.5.4.1 Free-Vibration Test There are several ways to input initial conditions for a free-vibration test. Initial nonzero displacement, initial nonzero velocity, or a combination of the two are possible initial inputs. To apply an initial displacement, the structure is first displaced by a tensioned cable. The cable is then released suddenly, causing the structure to vibrate freely. The natural period of the structure can be obtained from the time-versus-response curve. The damping coefficient can also be estimated by measuring the decrease in displacement amplitude as time passes. To input initial velocity into the structure, the structure can be hit by a swinging pendulum. Alternatively, a rocket attached to the structure can be propelled for a short time (Ohsaki, 1967; GDTHB, 1969).

2.5.4.2 Forced-Vibration Test The equipment required for a forced-vibration test is more elaborate than that for a free-vibration test. The time required accordingly is longer, but the data obtained are more accurate and useful.

The basic procedure of forced-vibration testing is the following:

1. Drive an excitation generator with a constant frequency until the test structure develops a steady-state vibration.

2. Measure the displacement response.

3. Change the frequency and repeat steps 1 and 2.

4. Plot the measured displacements with respect to the frequency.

Figure 2-40 shows an example of such a curve (GDTHB, 1969). The fundamental natural frequency can be found instantly from this figure. Further, the fraction of critical damping ξ can be estimated from the peak width of the curve as

$$\xi = \frac{1}{2} \frac{\Delta f}{f_n} \tag{2-169}$$

where f_n = resonant frequency
   Δf = width of the resonance curve at the amplitude level of $1/\sqrt{2}$ of the peak amplitude

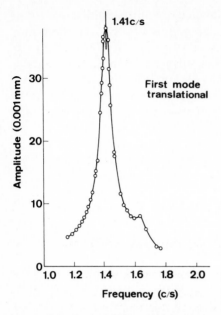

Figure 2-40 Resonance curve by GDTHB. [*GDTHB, Summarized report on dynamic tests of high-rised buildings and cooperative plan for large-scale vibration tests in Japan*, Proc. Fourth World Conf. Earthquake Eng., *Santiago*, *1*, *B1-111–125 (1969)*.]

By suddenly stopping the motion of the excitation generator, the test structure is forced into free vibration. The natural frequency and damping of the structure can also be found in this way.

Instead of being changed discretely, the frequency of an excitation generator may be changed continuously from low to high and back to low. The frequency is increased continuously until the amplitude reaches its peak value and starts to decrease. The generator is then stopped or, alternatively, the frequency is decreased, and the vibration amplitude of the structure is continuously measured. This procedure, named the *run-down test*, is simple and needs little time to obtain the resonance curve. The curve, however, is not accurate enough to allow the damping ratio to be obtained accurately. The test should therefore be used for preliminary observation of the resonance condition of the structure. For accurate results, the discrete and more precise procedure of forced-vibration testing should be carried out.

The excitation of a generator is usually achieved by rotating an eccentric mass installed in the generator. Let us suppose that the eccentric mass, radius, and angular frequency of a pendulum are m, r, and ω, respectively, and the induced inertial force is $mr\omega^2$. The force component in one direction is calculated as $mr\omega^2 \sin \omega t$. With two such pendulums paired as shown in Fig. 2-41, a sine wave in one direction only can be generated.

In another type of excitation generator a mass moves forward and backward alternately, with the driving power provided by a compressor, magnet, or hydraulic jack.

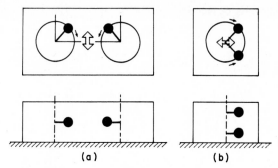

Figure 2-41   Eccentric-weight vibration generators.

In some tests, many generators are installed in various parts of the building structure, and the vibrations are synchronized (Hudson, 1964; GDTHB, 1969; Hudson, 1970).

Useful data on the dynamic characteristics of building structures can also be obtained by measuring the strong motion of the structures with the use of the SMAC or by observing microtremors caused by small earthquake disturbance or vibration caused by vehicles (GDTHB, 1969).

2.5.4.3 Shaking-Table Test   The shaking-table test is considered to be the most direct means of simulating the earthquake behavior of a structure. To study the seismic safety of various types of equipment in a building structure or a nuclear reactor, a shaking-table test is often conducted at full scale. Some test laboratories have huge shaking tables which can accommodate the large masses involved. In most cases, however, scale models of structures are tested on shaking tables. Many existing shaking tables are between 3 m and 5 m long, so that the size of the test building structures ranges at most from one-fifth to one-hundredth of the prototype (Clough and Tang, 1975; Otani, 1975; Nakamura, Yoshida, et al., 1980). In fabricating scale models, the scale effect needs special consideration (Hudson, 1961). To preserve the prototype relationship, materials used in scale models sometimes differ from those of the prototypes. In general, however, reinforced-concrete or steel scale models are fabricated for reinforced-concrete or steel prototype building structures.

Shaking tables are designed to be driven in one or two horizontal directions or in combined horizontal and vertical directions. The input motion can be sine waves, triangular waves, arbitrary generated waves, or motions recorded in real earthquake disturbances. The capacity of a shaking table is defined in terms of dynamic characteristics, table dimensions, maximum carrying load, maximum acceleration, and other properties. The driving power is either electrohydraulic or electromagnetic.

2.6 Inelastic-Response Analysis of Structures

2.6.1 Significance of Inelastic-Response Analysis

Discussion so far has been devoted to linear elastic systems. The inelastic behavior of structures is as important as elastic behavior in seismic design practice for the following reasons. A building structure should perform without damage under the small to medium earthquakes which will occur occasionally during its lifetime. In addition, it should not collapse under strong earthquake motion with a recurrence time of 50 years or more. A structure designed with this philosophy is often subjected to medium earthquake forces which bring it into the inelastic range. On some occasions, observed forces have been 3 to 4 times as large as those specified in codes. Nevertheless, the structures were undamaged in most cases. Energy dissipation due to hysteretic damping is believed to be an additional safety margin possessed by these structures. To design structures so that they remain elastic under large earthquake motions is very uneconomical and is considered to be unrealistic except for brittle masonry structures with large lateral stiffness. The effect of energy dissipation caused by the hysteretic behavior of a building structure should therefore be evaluated accurately from an inelastic analysis of the structure.

Elastic analysis with use of the concept of equivalent viscous damping is one way to evaluate the effect described in Sec. 2.5.2.1. This analytic procedure is liable to large error when the ductility factor is large. For a more accurate evaluation, plastic deflections may be estimated on the basis of the corresponding elastic deflection in Sec. 2.6.3. Alternatively, a proper elastic-inelastic analysis may be performed.

2.6.2 Methods of Nonlinear-Response Analysis

The mode-superposition technique is useful in elastic analysis but is not applicable in inelastic analysis because the principle of superposition is no longer valid. The most popular analytic method for inelastic systems is the step-by-step direct-integration method in which the time domain is discretized into many small intervals Δt and for each time interval the equations of motion are solved with the displacements and velocities at the previous step serving as initial data. The stiffness characteristics at the outset of the considered time interval are assumed to be constant throughout this step. Computation proceeds in a step-by-step manner. The following is a brief example of this procedure, but readers may wish to refer to Clough and Penzien (1975) for more details. Let us consider the one-mass system shown in Fig. 2-1 and assume that its hysteresis loop is similar to those in Figs. 2-35 through 2-37. Here the spring constant changes with time. The equation of motion [Eq. (2-1)] should be satisfied at all time steps. Subtracting the equation of motion at time t from the equation at time $t + \Delta t$ gives an incremental equation of motion as

$$\Delta F_I(t) + \Delta F_D(t) + \Delta F_s(t) = \Delta F(t) \tag{2-170}$$

This equation can also be written in the form of Eq. (2-3). Thus,

$$m\Delta \ddot{v}(t) + c\Delta \dot{v}(t) + k(t)\Delta v(t) = \Delta F(t) \tag{2-171}$$

Here it is best to employ slope AB of Fig. 2-42 for the stiffness $k(t)$. This slope, however, cannot be estimated precisely in advance.

In most cases, slope AC, which is the tangential slope at point A, is used to represent $k(t)$. Various integration schemes have been proposed for solving Eq. (2-171). Here the method in Clough and Penzien (1975) is explained. Let us assume that the acceleration change is linear during the time increment Δt and that the acceleration is $\ddot{v}(t) + \Delta \ddot{v}(t)$ at the end of the increment. According to the relationship between acceleration, velocity, and displacement, velocity and displacement change, respectively, in a parabolic and a cubic polynomial form. Let us consider the velocity and displacement to be $\dot{v}(t) + \Delta \dot{v}(t)$ and $v(t) + \Delta v(t)$ at the end of the time increment. From these equations, $\Delta \ddot{v}(t)$ and $\Delta \dot{v}(t)$ can be expressed in terms of $\Delta v(t)$. By substituting $\Delta \ddot{v}(t)$ and $\Delta \dot{v}(t)$ into Eq. (2-171), we can obtain

$$\tilde{k}(t)\Delta v(t) = \Delta \tilde{F}(t) \tag{2-172}$$

in which $\tilde{k}(t)$ and $\Delta \tilde{F}(t)$ are the effective stiffness and effective load increment, respectively. In this equation, $\tilde{k}(t)$ and $\Delta \tilde{F}(t)$ are known values, and so $\Delta v(t)$ can be solved. $\Delta \dot{v}(t)$ can also be solved by use of $\Delta v(t)$. It is now possible to set up the initial conditions for the time increment: $\dot{v}(t) + \Delta \dot{v}(t)$ and $v(t) + \Delta v(t)$. The initial value of $\ddot{v}$ at time t is given by

$$\ddot{v}(t) = \frac{1}{m} [F(t) - F_D(t) - F_s(t)] \tag{2-173}$$

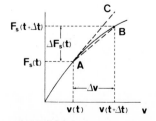

Figure 2-42　Representation of nonlinear stiffness.

The integration procedure can thus be summarized as follows:

1. Initial data at time t, i.e., $\dot{v}(t)$ and $v(t)$, are given from the computation of the preceding step.

2. $k(t)$ is taken to be the tangential stiffness at point A in Fig. 2-42.

3. $\ddot{v}(t)$ is solved from Eq. (2-173), and $\tilde{k}(t)$ and $\Delta\tilde{F}(t)$ are computed. $\Delta v(t)$ is obtained by solving Eq. (2-172), and finally $\Delta\dot{v}(t)$ is calculated.

4. The velocity and displacement at the end of the time step are computed by adding the previously determined increments, $\Delta\dot{v}(t)$ and $\Delta v(t)$, to $\dot{v}(t)$ and $v(t)$. These now serve as the initial conditions for the next step.

This procedure can readily be extended to MDOF systems. For an MDOF system, an expression similar to Eq. (2-171) is given as

$$\mathbf{m}\Delta\ddot{\mathbf{v}}(t) + \mathbf{c}\Delta\dot{\mathbf{v}}(t) + \mathbf{k}(t)\Delta\mathbf{v}(t) = \Delta\mathbf{F}(t) \tag{2-174}$$

The integration procedure is identical with that described above.

The accuracy of the results increases with a smaller time interval. It has been verified that a time interval smaller than one-tenth of the response period provides sufficient accuracy and that a time interval larger than a certain limiting value leads to computational instability. In most cases, the interval is selected so that the ground acceleration is reasonably simulated (Clough and Penzien, 1975). The Runge-Kutta method is also used frequently for direct integration; it provides more accurate results than the linear-acceleration method.

2.6.3 Inelastic-Response Behavior

Let us consider a one-mass system which is either linear elastic and perfectly plastic or bilinear with a positive second branch. The relationship between the maximum elastic response—the response of the system when it remains elastic regardless of the force level—and the maximum inelastic response can be estimated in the following manner (Veletsos and Newmark, 1960; Osawa and Shibata, 1961).   Figure 2-43 shows the maximum inelastic response of one-mass systems which have an identical natural period and various levels of yield force. Figure 2-43a is for the case in which the natural period of the systems is in a constant-velocity range as in Fig. 2-13. From this figure, the maximum inelastic response δ_p is found in such a way that the elastic response is A, the yield level is B, line BC is parallel to the abscissa, C' is the intersection of BC with the chain line, and the δ value of C' is δ_p. As is evident from this figure, the maximum inelastic displacement response is nearly constant regardless of the level of yield force except for the range in which the yield level is

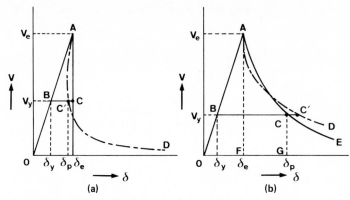

Figure 2-43 Relationship between elastic and inelastic responses. *(a)* System with long natural vibration period (S_v = const). *(b)* System with short natural vibration period (S_a = const).

extremely low. In this range, the displacement response increases precipitously. The ratio of the maximum displacement δ_p to the yield displacement δ_y is expressed as

$$\mu = \frac{\delta_p}{\delta_y} \tag{2-175}$$

The value μ is defined as the *ductility ratio*. As long as the yield force is not extremely low, $\delta_p = \delta_e$ and, therefore,

$$V_y = \frac{V_e}{\mu} \tag{2-176}$$

From this relationship, it can be stated that a system whose maximum elastic response is V_e and whose ductility ratio is μ would safely sustain the earthquake disturbance if its yield-force level is more than V_y. Alternatively, it can be said that a system whose elastic response is V_e and whose yield-force level is V_y needs to have a ductility ratio of μ,

$$\mu = \frac{V_e}{V_y} \tag{2-177}$$

for the system not to fail under the assumed earthquake disturbances (Clough and Penzien, 1975).

Figure 2-43*b* shows another pattern, in which response falls in a constant-acceleration spectrum as shown in Fig. 2-13. In this figure, C is the point at which the energy dissipated by a system having a yield force of V_y is equated with the energy to be dissipated by the system if the system is assumed to be perfectly elastic, while C' on the chain line shows the true displacement response of the system. Area □OBCG is thus taken to be equal to area △OAF. It has been verified that line *ACE* obtained in this manner reasonably duplicates the true response line *AC'D*. Thus, the following relationships can be written:

$$V_y = \frac{V_e}{(2\mu - 1)^{\frac{1}{2}}} \tag{2-178}$$

or $\qquad \delta_p = \dfrac{\mu}{(2\mu - 1)^{\frac{1}{2}}} \delta_e \tag{2-179}$

or $\qquad \mu = \dfrac{1}{2}\left[\left(\dfrac{V_e}{V_y}\right)^2 + 1\right] \tag{2-180}$

These equations, however, are hardly applicable if the second branch of a bilinear system has a negative slope. In this case, the response displacement is often biased in one direction, resulting in an instability. Prediction of inelastic deflection by these equations is also debatable for reinforced-concrete structures when their hysteresis curves are of the degrading type. Nevertheless, Eqs. (2-178), (2-179), and (2-180) are often used in design practice for the sake of simplicity.

The above procedure can be extended to MDOF systems. If the apparent ductility ratio

$$\mu_{eq_i} = \frac{\delta_{e_i}}{\delta_{y_i}} \tag{2-181}$$

is constant for all story levels, the relationship for SDOF systems [Eqs. (2-178) to (2-180)] can be approximately employed for each story. However, if μ_{eq_i} at one story is much larger than those at other stories, plastic deflection is concentrated at that story. The procedure of representing each story by an SDOF system is named the *ductility method* and is frequently adopted in design codes.

2.7 Measures of Aseismic Safety

2.7.1 Input Energy and Resisting Force

For the one-mass system shown in Fig. 2-2, the equation of motion when the spring is inelastic can be written as

$$m\ddot{v} + c\dot{v} + F_s(v) = F(t) \tag{2-182}$$

Multiplying by $\dot{v}$ on both sides of the equation and integrating from time t_1 and t_2 leads to

$$\tfrac{1}{2}m\dot{v}_2^2 - \tfrac{1}{2}m\dot{v}_1^2 = \int_{t_1}^{t_2} F(t)\dot{v}\,dt - \int_{t_1}^{t_2} c\dot{v}^2\,dt - \int_{t_1}^{t_2} F_s(t)\dot{v}\,dt \qquad (2\text{-}183)$$

The terms v_1 and v_2 are the values of v at point t_1 and t_2, respectively. Let us consider that the system governed by Eq. (2-183) is subjected to steady-state vibration as shown in Fig. 2-44. When v at point A is v_1 and v at point A after a single complete cycle is v_2, $v_1 = v_2$. The left-hand side of Eq. (2-183) then becomes zero. This indicates that the first term on the right-hand side, which represents the external work done during the cycle, equals the sum of the energy dissipated by viscous damping and the energy dissipated by hysteretic damping. It is clear that the larger the area of the hysteresis loop, the more input the system can resist.

Next, let us consider that the system is subjected to non-steady-state ground motion, with $t_1 = 0$ at the outset of the motion and t_2 at the end of the motion. In this condition, Eq. (2-183) can be explained as follows: The first term on the left-hand side $m\dot{v}_2^2/2$ is the kinetic energy at the end of the motion; the first term on the right-hand side, the input energy E; the second term, the energy dissipated by viscous damping; and the third term, the sum of the energy dissipated by hysteretic damping and the elastic-strain energy stored in the system. Thus,

$$W_p + W_e + W_D = E \qquad (2\text{-}184)$$

W_p is the accumulated energy dissipation by hysteretic damping and is considered to be a reasonable index for evaluating structural damage (Kato and Akiyama, 1975). Housner (1956, 1959) suggested the use of the velocity spectrum to define the intensity of ground motion. According to his suggestion, the system will be safe if

$$W_p + W_e \leqq E_H \qquad (2\text{-}185)$$

when no viscous damping exists. In this equation, E_H is the energy associated with the structural damage and is given as

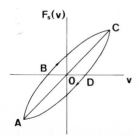

Figure 2-44 Load-deflec-tion hysteresis loop.

$$E_H = \frac{mS_v^2}{2} = \frac{m\dot{v}_{max}^2}{2} \qquad\qquad (2\text{-}186)$$

2.7.2 Global and Local Ductility Factors

As discussed in Sec. 2.6.3, structures that can sustain large plastic deformation have good earthquake resistance. Accordingly, structures having large μ values can be designed with lower levels of lateral-force-resisting capacity. It can also be stated that the part of the structure which is to undergo large plastic deformation should possess a ductility factor that will more than cover the expected plastic deformation. The ductility factor defined in Sec. 2.6.3 is associated with horizontal story deflection and is determined on the assumption that the horizontal force-deflection relationship is linear elastic and perfectly plastic. However, when the deformation capacity of structural details is examined, another definition of ductility is often more appropriate. Based on work done by Mahin and Bertero (1976), various definitions of ductility factor will be introduced in the next section.

2.7.2.1 Ductility Factors for Overall Response

Displacement Ductility Factors. As discussed in Sec. 2.6.3, there are various proposals for determining the static lateral force in terms of the ductility factor. Some of these proposals have been incorporated into design practice. It is implicitly assumed in these proposals that yielding occurs simultaneously in all story levels, but this is not true on many occasions. Furthermore, the assumption that the horizontal force-deflection relationship is linear elastic and perfectly plastic does not hold in most cases since the plastic story deflection occurs gradually as plastic hinges form in the members. A typical horizontal force-deflection curve is illustrated as a dashed line in Fig. 2-45. In this curve, some idealization is required for determining the ductility factor. The yield deflecton δ_y can

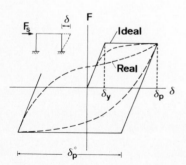

Figure 2-45 Definition of ductility factor.

be defined in many ways as shown in Fig. 2-46. Since plastic deflection capacity under load reversal differs from capacity when the load is continously applied in one direction, δ_p is often replaced by δ_p^* in Fig. 2-45. Another proposal is to use the accumulated plastic deflection which occurs during load reversal to allow for accumulative structural damage.

Drift Ductility Factors. In a high-rise building structure, total horizontal deflection is the sum of the story shear deflection and the lateral drift caused by bending of the structure as shown in Fig. 2-47. In this case, tangential drift, obtained by subtracting bending drift from total drift, can be used to check the performance of nonstructural elements. The drift is not nondimensionalized in most design conditions. Such a treatment must be made not for δ_y but for the failure drift of nonstructural elements if this is necessary for systematic judgment of the performance of these elements.

2.7.2.2 Ductility Factors for Critical Regions

Rotation Ductility Factors. Joint rotation or plastic-hinge rotation is an index for evaluating the plastic-deformation capacity of local regions. Such rotation can be determined by examining local buckling and frac-

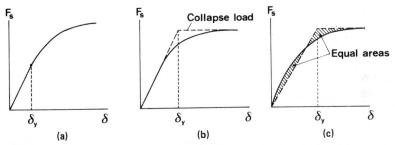

Figure 2-46 Definition of yield deformation. *(a)* Based on first yielding. *(b)* Based on initial yielding. *(c)* Based on energy absorption.

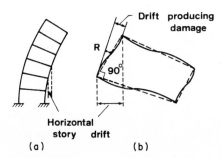

Figure 2-47 Story drift of a building structure. *(a)* Horizontal story drift. *(b)* Drift producing damage.

ture in joints for steel structures and the crushing of concrete at joint regions in reinforced-concrete structures. If a plastic hinge is assumed to have zero length, the definition of plastic rotation is rather ambiguous and difficult to quantify. In most cases, such plastic rotation is not nondimensionalized, and values obtained from experiments or otherwise are employed directly for judgment purposes.

Ductility Factors for Curvature. Since the moment-curvature relationship of a wide-flange steel member is elastic–perfectly plastic, computation of the ductility factor is straightforward. However, it should be noted that the curvature capacity is a function of the applied axial force in the steel member. In reinforced-concrete members, the ductility factor is difficult to define.

2.7.3　Effect of Deterioration

Under large earthquake motions, particularly those of long duration, building structures sustain many large load reversals. In highly seismic areas, building structures are also subject to occasional earthquake motions. For these cases, the effect of structural deterioration (due to previous load reversal) on the performance of the structure needs to be examined carefully when a new earthquake motion occurs.

In some types of structures, strength may be degraded after each complete load reversal (Fig. 2-48). Such structures apparently would

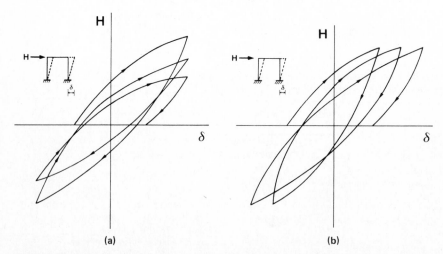

(a)　　　　　　　　　　　　　　　　　　(b)

Figure 2-48　Degradation by load reversal. *(a)* Strength degradation. *(b)* Stiffness degradation.

deform to a greater degree in succeeding cycles of loading than structures which do not exhibit any such degradation. Details of degradation will be discussed in Chap. 3.

Steel is ductile and reveals very stable behavior under load reversal. Steel members in frames, however, may not be so ductile in certain conditions even though the material is ductile. The deterioration of such members and frames occurs primarily by buckling or brittle failure of connections. As an example, a frame whose columns undergo local buckling at the ends displays a degrading hysteresis as shown in Fig. 2-48.

The degradation behavior of concrete structures is more significant than that of steel members. Degradation is noticeable particularly when shear or bond failure takes place in members, connections, or shear walls. It is not surprising that in some cases load-carrying capacity is halved after only several cycles of load reversal.

The hysteretic behavior of composite structures is a combination of those of steel and reinforced-concrete structures. Buckling, which often occurs in steel structures, is restrained by the concrete surrounding the steel components. The major cause of degradation is attributed to the characteristics of the reinforced-concrete components. It can thus be stated that composite structures with greater proportions of steel display better hysteretic performance.

In design practice degradation is undesirable, and details must therefore be chosen as carefully as possible to avoid such degradation. In turn, if degradation is inevitable, analysis and design should allow for it. A degrading model indicated in Fig. 2-36 is often insufficient to represent the true degrading mechanism, but very few models which allow for drastic degradation in hysteresis have yet been proposed.

2.7.4 Criteria of Failure

The ultimate safety of structures cannot be judged without explicit criteria for structural failure. Brittle structures, such as masonry structures which do not contain reinforcement, fail when the maximum applied force exceeds the strength of the structure. Such cases are clear-cut. On the other hand, steel structures do not fail even when stresses due to external earthquake forces reach yield limits; they are safe as long as the plastic deformation, and hence the ductility ratio, is within allowable limits. There is a hypothesis that the allowable ductility ratio is related to low cycle fatigue of the structure (Minai, 1970; Yamada, 1972; Yamada and Kawamura, 1976). *Low cycle fatigue* is defined as failure caused by load reversal in which the force level is close to the yield force for the structure. This concept implies that damage will not accumulate if the applied force is below a certain level. Kato and Akiyama (1977) have proposed another hypothesis in which the term *cumulative plastic deflection* is introduced. In this concept, failure is taken to occur when the cumulative plastic deflection caused by load reversal reaches a limit value.

REFERENCES

Aoyama, H. (1980). Trends in the earthquake resistant design of SRC highrise buildings, in B. Kato and L.-W. Lu (eds.), *Developments in Composite and Mixed Construction*, Proc. U.S.A.–Japan Seminar on Composite Structures and Mixed Construction Systems, Gihodo Shuppan Co., Tokyo, 171–181.

Cherry, S. (1974). Estimating underground motions from surface accelerograms, in J. Solnes (ed.), *Engineering Seismology and Earthquake Engineering*, Noordhoff International Publishing, Leiden, 125–150.

Clough, R. W., and S. B. Johnston (1966). Effect of stiffness degradation on earthquake ductility requirements, *Proc. Second Japan Earthquake Eng. Symp.*, Tokyo, 227–232.

―――― and J. Penzien (1975). *Dynamics of Structures*, McGraw-Hill Book Company, New York.

―――― and D. Tang (1975). Seismic response of a steel building frame, *Proc. U.S. Nat. Conf. Earthquake Eng.*, Ann Arbor, Mich., 268–277.

Dowrick, D. J. (1977). *Earthquake Resistant Design*, John Wiley & Sons, Inc., New York.

GDTHB (The Group for Dynamic Tests of High-Rised Buildings) (1969). Summarized report on dynamic tests of high-rised buildings and cooperative plan for large-scale vibration tests in Japan, *Proc. Fourth World Conf. Earthquake Eng.*, Santiago, **1**, B1-111–125.

Housner, G. W. (1956). Limit design of structures to resist earthquakes, *World Conf. Earthquake Eng.*, San Francisco, 5-1–13.

―――― (1959). Behavior of structures during earthquakes, *J. Eng. Mech. Div., Am. Soc. Civ. Eng.*, **85**(EM-4), 109–129.

―――― (1970). Strong ground motion, in R. L. Wiegel (ed.), *Earthquake Engineering*, Prentice-Hall, Inc., Englewood Cliffs, N.J., 75–91.

―――― and A. G. Brady (1963). Natural periods of vibration of buildings, *J. Eng. Mech. Div., Am. Soc. Civ. Eng.*, **89**(EM-4), 31–65.

Hudson, D. E. (1961). Scale-model principles, in C. M. Harris and C. E. Crede (eds.), *Shock and Vibration Handbook*, McGraw-Hill Book Company, New York, vol. 2, chap. 27, 1–18.

―――― (1964). Resonance testing of full scale structures, *J. Eng. Mech. Div., Am. Soc. Civ. Eng.*, **90**(EM-3), 1–19.

―――― (1970). Dynamic tests of full-scale structures, in R. L. Wiegel (ed.), *Earthquake Engineering*, Prentice-Hall, Inc., Englewood Cliffs, N.J., 127–149.

Iwan, W. D. (1965). The steady state response of the double bilinear hysteretic model, *Trans. Am. Soc. Mech. Eng., J. Appl. Mech.*, **32**, 921–925.

Jacobsen, L. S. (1930). Steady forced vibration as influenced by damping, *Trans. Am. Soc. Mech. Eng.*, **52**(22), part 1, 169.

Jennings, P. C. (1964). Periodic response of a general yielding structure, *J. Eng. Mech. Div., Am. Soc. Civ. Eng.*, **90**(EM-2), 131–165.

―――― (1968). Equivalent viscous damping for yielding structures, *J. Eng. Mech. Div., Am. Soc. Civ. Eng.*, **94**(EM-1), 103–116.

Kanai, K. (1962). On the period and the damping of vibration in actual buildings, *Bull. Earthquake Res. Inst.*, University of Tokyo, **39**, part 3.

Katayama, T. (1969). Damping characteristics of building structures, *Trans. Arch. Inst. Japan*, no. 157, 11–18 (in Japanese).

Kato, B., and H. Akiyama (1975). Energy input and damages in structures to severe earthquakes, *Trans. Arch. Inst. Japan*, no. 235, 9–18 (in Japanese).

—— and —— (1977). Earthquake resistant design for steel buildings, *Proc. Sixth World Conf. Earthquake Eng.*, New Delhi, **2**, 1945–1950.

Mahin, S. A., and V. V. Bertero (1976). Problems in establishing and predicting ductility in seismic design, *Proc. Int. Symp. Earthquake Struct. Eng.*, University of Missouri–Rolla, St. Louis, **1**, 613–628.

Minai, R. (1970). On the aseismic safety of building structures, *Annals: Disaster Prev. Res. Inst.*, Kyoto University, no. 13A, 5–22 (in Japanese).

Müller, F. P., and E. Keintzel (1978). *Erdbebensicherung von Hochbauten*, Verlag von Wilhelm Ernst & Sohn, Berlin.

Nakagawa, K. (1960). Vibration characteristics of reinforced concrete buildings existing in Japan, *Proc. Second World Conf. Earthquake Eng.*, Tokyo, **2**, 973–982.

Nakamura, T., N. Yoshida, S. Iwai, and H. Takai (1980). Shaking table test of steel frames, *Proc. Seventh World Conf. Earthquake Eng.*, Istanbul, **7**, 165–172.

Newmark, N. M., and E. Rosenblueth (1971). *Fundamentals of Earthquake Engineering*, Prentice-Hall, Inc., Englewood Cliffs, N.J.

Ohsaki, Y. (1967). The use of jet reaction for dynamic tests of buildings, *Trans. Arch. Inst. Japan.* no. 142, 9–14.

Osawa, Y., and A. Shibata (1961). Characteristic of the nonlinear response of one mass systems under earthquake excitations, *Trans. Arch. Inst. Japan*, no. 69, 401–404 (in Japanese).

Otani, S. (1975). Earthquake tests of shear wall-frame structures, *Proc. U.S. Nat. Conf. Earthquake Eng.*, Ann Arbor, Mich., 278–286.

Otsuki, Y. (1960). Design seismic forces for reinforced concrete buildings, *Proc. Second World Conf. Earthquake Eng.*, Tokyo, **3**, 1947–1962.

Rea, D., R. W. Clough, J. G. Boukamp, and U. Vogel (1969). Damping capacity of a model steel structure, *Proc. Fourth World Conf. Earthquake Eng.*, Santiago, **1**(B-2), 63–73.

San Francisco ASCE-SA (Joint Committee of the San Francisco, California, Section, ASCE, and the Structural Associations) (1952). Lateral forces of earthquake and wind, *Proc. Am. Soc. Civ. Eng.*, **77**(66).

Shiga, T. (1976). *Vibration of Structures*, Kyoritsu Shuppan Co., Tokyo (in Japanese).

Tajimi, H. (1965). *Vibration of Buildings*, Corona Publishing Co., Tokyo (in Japanese).

Takeda, T., M. A. Sozen, and N. N. Nielsen (1970). Reinforced concrete response to simulated earthquakes, *J. Struct. Div., Am. Soc. Civ. Eng.*, **96**(ST-12), 2557–2573.

Tanabashi, R., and K. Kaneta (1962). On the relation between the restoring force characteristics of structures and the pattern of earthquake ground motions, *Proc. First Japan Earthquake Eng. Symp.*, Tokyo, 57–62.

Taniguchi, T. (1960). Seismic wall effect in framed structure in relation of the period of tall buildings, *Proc. Second World Conf. Earthquake Eng.*, Tokyo, **2**, 1013–1028.

Toki, K. (1981). *Earthquake Resistant Analysis of Structures*, Gihodo Shuppan Co., Tokyo (in Japanese).

Umemura, H. (1962). A brief consideration of the tall building seismic coefficient, *Trans. Arch. Inst. Japan*, no. 72, 19–23 (in Japanese).

Veletsos, A. S., and V. V. D. Nair (1975). Seismic interaction of structures on hysteretic foundations, *J. Struct. Div., Am. Soc. Civ. Eng.*, **101**(ST-1), 109–129.

—— and N. M. Newmark (1960). Effect of inelastic behavior on the response of simple systems to earthquake motions, *Proc. Second World Conf. Earthquake Eng.*, Tokyo, **2**, 895–912.

Yamada, M. (1972). Effect of cyclic loading on buildings, *Proc. Int. Conf. Planning Des. Tall Build.*, Lehigh University, Bethlehem, Pa., **2**(TC-18), 725–740.

—— and H. Kawamura (1976). Ultimate state aseismic design method, *Trans. Arch. Inst. Japan*, no. 240, 39–50.

3

BEHAVIOR OF BUILDING STRUCTURES UNDER EARTHQUAKE LOADING

3.1 Introduction

Good earthquake-resistant design requires a deep knowledge of how structures behave under earthquake loading. A structural design which merely meets code requirements is not satisfactory. From this point of view, the behavior of structures of different types is discussed in detail in this chapter.

As described in Sec. 2.7 in relation to the dynamic response of structures, several factors affect the performance of a structure in an earthquake. Figure 3-1 illustrates the horizontal load-displacement relationships of two different frames. The frame in Fig. 3-1a shows poor earthquake-resistant capacity: Strength deteriorates after displacement exceeds the value corresponding to maximum strength, and hence ductility is small; the hysteresis loops are of pinched shape, and the area enclosed by a hysteresis loop, which represents energy-dissipation capacity, is small. The strength degrades owing to repetition of loading. On the other hand, the frame in Fig. 3-1b shows good capacity: large ductility, large energy-dissipation capacity, and stable hysteresis loops without strength degradation (Wakabayashi, 1973).

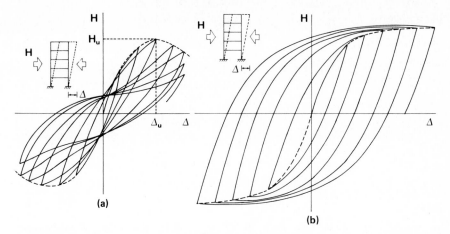

Figure 3-1   Behavior of structures under repeated horizontal load. *(a)* Poor behavior. *(b)* Good behavior.

In the following sections, ultimate strength, ductility, and behavior under repeated loading of materials, members, connections, and systems are described. Methods of improving earthquake-resistant capacity are also discussed. Some of the structural damage caused by past earthquakes is considered. The behavior of structures under earthquake loading has been reviewed in several references (Wakabayashi, 1972; Wakabayashi, 1977*b*; Tall Building Committee 15, 1979).

3.2 Behavior of Construction Materials

3.2.1 Concrete

The compressive strength of concrete is usually obtained by test from a cylinder 30 cm high by 15 cm in diameter or, in some countries, from a cube with 20-cm sides. Figure 3-2*a* shows parabolic-shaped stress-strain curves for concrete cylinders. Young's modulus E_c may be taken to be equal to $4730\sqrt{f_c'}$ MPa, where f' designates compressive strength, MPa (ACI Committee 318, 1983*a*, 1983*b*). The value of strain at maximum stress is approximately 0.002 regardless of strength. The shape of the descending branch of the stress-strain curve varies with the amount of transverse reinforcement, which provides a confining effect (see Sec. 3.2.2).

It is necessary to develop an idealized model of the stress-strain relationship in order to simplify the calculation of flexural strength and

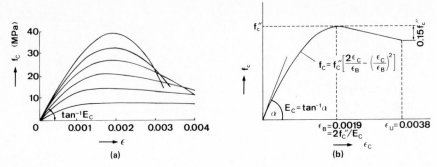

Figure 3-2 Stress-strain relationships of concrete under compressive loading. *(a)* Actual relationship. *(b)* Idealized relationship.

deformation of members. Figure 3-2*b* shows a simple model proposed by Hognestad (1952), which consists of a parabola and a straight line. Figure 3-3*a* shows an experimentally obtained stress-strain relation under repeated loading, from which the idealized model shown in Fig. 3-3*b* is derived (Blakeley and Park, 1973). A detailed survey of models of the

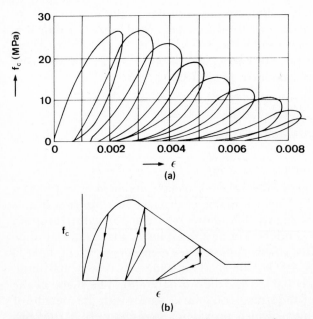

Figure 3-3 Stress-strain relationships of concrete under repeated compressive loading. *(a)* Actual relationship. *(b)* Idealized relationship. [*From R. W. G. Blakeley and R. Park, Prestressed concrete sections with cyclic flexure,* J. Struct. Div., Am. Soc. Civ. Eng., *99(ST-8), 1717–1742 (1973).*]

concrete stress-strain relationship is found in Aoyama (1981). The modulus of rupture of concrete, obtained from the flexural test, is given conservatively by the formula $0.62\sqrt{f_c'}$ MPa (ACI Committee 318, 1983). The tensile strength of concrete is measured by a test in which a cylinder placed horizontally and loaded along a diameter splits. Its value lies between 50 and 75 percent of the modulus of rupture.

When concrete is subjected to high-speed compression, as in earthquake loading, f_c-ε curves become as shown in Fig. 3-4. Although the shape of the curve is unchanged, the trend is for maximum strength to increase with increasing strain rate. For example, maximum stress for a strain rate of 0.5 percent per second is 14 percent higher than that obtained by the quasi-static test (Wakabayashi, Nakamura, et al., 1978).

3.2.2 Steel

The stress-strain relationship for steel, shown in Fig. 3-5a, is usually idealized to the bilinear form shown by solid lines in Fig. 3-5b, although strain hardening (dashed line) is taken into account in some cases. F_y and F_u are used for steel sections or plates, and f_y is used for reinforcing bars. The value of Young's modulus E_s is about 0.20×10^6 MPa.

For a structure to possess sufficient ductility, component material must be such that total elongation up to breaking failure is sufficiently large and the ratio of yield stress F_y to ultimate stress F_u is not close to unity. The latter requirement prevents the situation in which a tension member with bolt holes breaks on a net section before yielding takes place in a gross section.

In the case of steels such as prestressing steel, of which the stress-strain relationship does not clearly show the yield plateau, yield stress is defined as the stress which leaves the specimen with a permanent plastic strain of 0.2 percent when the specimen is unloaded.

The hysteretic stress-strain relationship for steel subjected to alter-

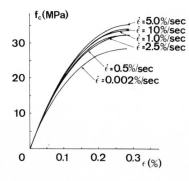

Figure 3-4 Effect of the strain rate on the stress-strain relationships of concrete. [*From M. Wakabayashi, T. Nakamura, et al., Effect of strain rate on stress-strain relationships of concrete and steel*, Proc. Fifth Japan Earthquake Eng. Symp., Tokyo, 1313–1320 (1978).*]

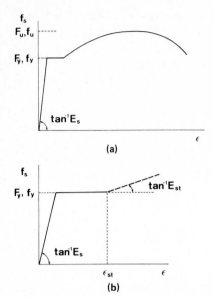

Figure 3-5 Stress-strain relationships of steel. *(a)* Actual relationship. *(b)* Idealized relationship.

nately repeated loading is shown in Fig. 3-6*a*. The unloading branch shows an incipient slope equal to the elastic slope and is gradually softened owing to the Bauschinger effect. Examples of simple models are given in Fig. 3-6*b*, *c*, and *d*.

High-speed loading tests show an increase in yield stress when compared with the results of the quasi-static test. Increases of 8 percent under a strain rate $\dot{\varepsilon}$ equal to 0.5 percent per second and of 17 percent for $\dot{\varepsilon}$ equal to 10 percent per second are typical. However, the increase in ultimate stress is not more than 3 percent (Wakabayashi, Nakamura, et al., 1978).

3.3 Behavior of Reinforced-Concrete Structures

3.3.1 Introduction

It has been shown by the San Francisco earthquake (1906) and the Kanto earthquake (1923) that reinforced-concrete structures are far better than masonry structures with regard to earthquake resistance. Since they are also economical, they have been constructed in earthquake-prone areas. While damage caused by earthquakes has been decreasing owing to the improvement of earthquake design codes, the following factors, which exclude defects due to poor fabrication and erection, still must be counted as potential causes of damage:

1. Inadequate story shear strength caused by too few columns and walls

2. Brittle shear failure of columns or beams

3. Brittle shear failure of columns which have been shortened by the supporting effect of nonstructural elements

4. Slip of anchored bars or shear failure of the joint block in beam-to-column connections

5. Brittle failure of single or coupled shear walls, particularly shear walls with openings

6. Torsion caused by the noncoincidence on the floor plan of the center of gravity and the center of stiffness

7. Concentration of damage at a specific story because of unequal distribution of the ratio of stiffness along the height

8. Separation of secondary members such as exterior walls because of poor connections

Factors 1, 5, 6, and 7 are related to structural planning, while factors 2, 3, 4, and 8 may be avoided by improving the detailing.

In addition to the monolithic in situ reinforced-concrete structures discussed above, there are precast-concrete structures (Tall Building Committee 21E, 1979) and prestressed-concrete (in situ or precast) struc-

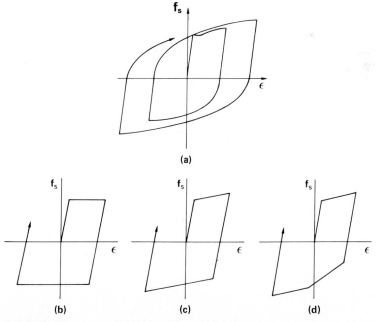

(a)

(b)　　　　　(c)　　　　　(d)

Figure 3-6 Hysteretic behavior of steel. (a) Real behavior. (b) Elastic plastic model. (c) Bilinear model. (d) Bauschinger model.

tures. Connections in a precast-concrete structure are fabricated at the construction site; hence its earthquake resistance may be poorer than that of a monolithic reinforced-concrete structure.

It is particularly important to provide sufficient strength, rigidity, and ductility of connection in the design of a precast-concrete structure. In comparison with a reinforced-concrete structure, a prestressed-concrete structure possesses small energy-dissipation capacity since it behaves rather elastically. Also high stresses can develop in the concrete around the anchorage of the prestressing steel. The design of prestressed-concrete structures must allow for these disadvantages with regard to earthquake-resistant capacity.

3.3.2   Interaction between Concrete and Steel

3.3.2.1   Bond between Reinforcing Bars and Concrete   Bond strength between round bars and concrete is provided by chemical adhesion and friction. Once slip occurs, a further bond can be developed only by friction. With deformed bars, bond strength at incipient slip is not much different from that of round bars, but resistance increases with the progress of the slip since the ribs are wedged into the concrete. When a deformed bar is embedded with sufficient cover in concrete which is transversely reinforced against splitting, the concrete between the ribs eventually crushes and the bar pulls out. In practical cases, however, pullout of the bar is often accompanied by splitting of the surrounding concrete. The bond strength associated with this failure mechanism rises with increasing thickness of the concrete cover and increasing transverse reinforcement.

The relationship between bond stress and slip of a deformed bar embedded in concrete and subjected to repeated loading becomes as shown in Fig. 3-7. The shape of the hysteresis loop is of the slip type, and strength degradation is observed (Morita and Kaku, 1973; Viwathanatepa, Popov, and Bertero, 1979).

3.3.2.2   Confining Effect of Transverse Reinforcement   When the stress in a concrete cylinder approaches compressive strength, internal cracking occurs progressively and the concrete expands transversely. If the compression zone is confined by transverse reinforcement such as spirals and hoop ties, ductility of the concrete is greatly improved, as shown in Fig. 3-8 (Muguruma, Watanabe, et al., 1979). If square hoop ties are used in the member, the concrete along the diagonals of the tie is confined, as shown in Fig. 3-9, and as a result stress-strain relationships for the confined concrete become as shown in Fig. 3-10, where it is observed that the

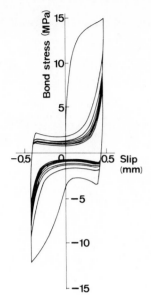

Figure 3-7 Bond-stress−slip relationship. [*From S. Morita and T. Kaku, Local bond stress-slip relationship under repeated loading,* Symposium on Resistance and Ultimate Deformability of Structures Acted on by Well-Defined Repeated Loads: Preliminary Report, *International Association of Bridge and Structural Engineering, Lisbon Symposium, 221−227 (1973).*]

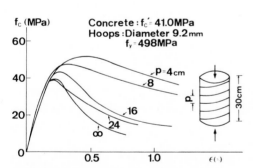

Figure 3-8 Effect of confinement by spiral reinforcement on stress-strain relationships.

Unconfined concrete

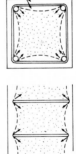

Figure 3-9 Confinement of concrete by square hoop ties.

slope of the descending branch of the curve is reduced by increasing amounts of the confining reinforcement (Kent and Park, 1971; Sheikh and Uzumeri, 1980; Sheikh, 1982).

3.3.2.3 Buckling of Reinforcing Bars

Longitudinal reinforcing bars under compression in beams and columns are prevented from buckling by the lateral restraint provided by concrete. However, when covering concrete subjected to high compressive stress becomes unstable, the restrain-

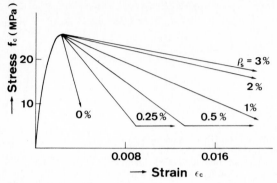

Figure 3-10 Influence of the quantity of hoops on the stress-strain curves of concrete in members. [*From D. C. Kent and R. Park, Flexural members with confined concrete*, J. Struct. Div., Am. Soc. Civ. Eng., **97**(ST-7), 1969–1990 (1971).]

ing effect is reduced and the bar buckles as shown in Fig. 3-11*a*, so that the axial force carried by the compression bar is reduced. This reduces the load-carrying capacity of the member. In order to minimize reduction of carrying capacity and to ensure sufficient ductility, it is necessary to set a limit to the effective length of the longitudinal reinforcing bar, i.e., the distance between the lateral supports provided by transverse reinforcement. Thus, code limits are placed on the ratio of the distance between transverse reinforcement to the diameter of the longitudinal reinforcing bar (ACI Committee 318, 1983*a*). The transverse reinforcement does not effectively support longitudinal reinforcing bars located at intermediate points between the corners, since the transverse reinforcement bends outward as shown in Fig. 3-11*b*. Diamond-shaped reinforcement as shown

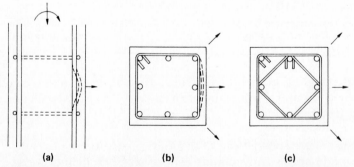

Figure 3-11 Effect of transverse reinforcement in preventing the buckling of the main reinforcement. (*a*) Buckling of the longitudinal reinforcement. (*b*) Without diamond ties. (*c*) With diamond ties.

in Fig. 3-11c is needed to make the effective length of such longitudinal bars equal to the distance between the transverse reinforcement. This type of reinforcement is effective for confining concrete and increasing maximum strength and failure strain of concrete (Scott, Park, and Priestley, 1982).

3.3.3. Flexural Behavior of Members

3.3.3.1 Load-Deformation Relationships   Figure 3-12 shows a moment (M)-curvature (φ) relationship for a doubly reinforced concrete beam in which steel content is below the balanced-reinforcement ratio. The curve is linear up to the incipient cracking point A, after which stiffness decreases. At point B, the tension steel yields, but strength still increases gradually and reaches its maximum value at point C. The compression concrete crushes at point D, the compression reinforcement buckles at point E, and strength then decreases rapidly.

Moment-curvature relationships for reinforced-concrete beam columns subjected to constant axial force and monotonically increasing curvature are shown in Fig. 3-13, where N_0 designates the maximum compressive strength of the column. The behavior for $N/N_0 = 0$ becomes

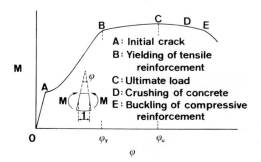

Figure 3-12 Moment-curvature relationship of a beam under monotonic loading.

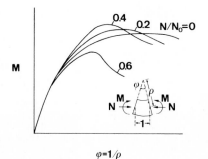

Figure 3-13 Moment-curvature relationships of members under constant axial force and monotonic bending moment.

the same as in Fig. 3-12 at the higher-moment area. Bending strength increases with increasing axial force, provided that the value of N/N_0 is below a certain value. The strength decreases as the axial-force ratio N/N_0 exceeds this value. On the other hand, ductility decreases with increasing axial force.

Figure 3-14 shows hysteretic moment (M)-rotation (R) relationships for beam columns which fail in flexure when subjected to constant axial force and alternately repeated antisymmetrical bending moments with shear (Wakabayashi and Minami, 1976). Two cycles of repeated loading are applied for each curvature amplitude. It is observed in the relationship for $N/N_0 = 0$ that the hysteresis loops are spindle-shaped, with some pinching due to the shear effect. Ductility and energy-dissipation capacity are large, and degradation of strength due to repetition of loading is small. The strength of the beam column with $N/N_0 = 0.3$ decreases at a relatively early stage after maximum strength has been attained, and a very brittle failure is observed in the case of $N/N_0 = 0.6$.

Figure 3-15 illustrates a hysteresis loop in the moment-curvature relationship for a reinforced-concrete element which is subjected to zero or very small axial force. Between points B and E, the cracks penetrate the whole section and do not close; hence the bending moment is carried only by the steel in this portion of the loop. This is the reason why the loop has a pinched shape. The portion between points C and D is softened owing to the Bauschinger effect.

3.3.3.2 Flexural Strength of Beams   The ultimate strength of a doubly reinforced concrete beam in flexure is assumed to be reached when the extreme fiber strain in the compression concrete becomes 0.003 as shown in Fig. 3-16. In the calculation, the actual shape of the concrete compressive-stress block is replaced by a rectangle shown in Fig. 3-16c,

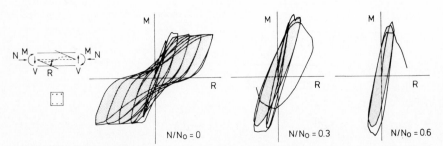

Figure 3-14   Relationships between moment and rotation angle for beam columns failing in flexure under constant axial force and alternately repeated antisymmetrical bending moment. [*From M. Wakabayashi and K. Minami, Experimental studies on hysteretic characteristics of steel reinforced concrete columns and frames*, Proc. Int. Symp. Earthquake Struct. Eng., *University of Missouri–Rolla, St. Louis*, **1**,467–480 (1976).]

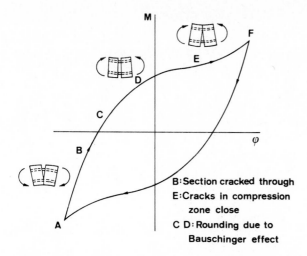

B: Section cracked through
E: Cracks in compression
    zone close
C D: Rounding due to
    Bauschinger effect

Figure 3-15   Moment-curvature relationship of a doubly reinforced concrete element under alternately repeated bending moments.

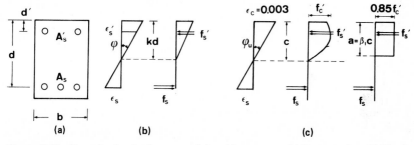

Figure 3-16   Beam in the elastic state and the ultimate state. *(a)* Cross section. *(b)* Elastic state. *(c)* Ultimate state.

where $\beta_1 = 0.85$ for concrete strength $f'_c \leq 27.6$ MPa, and the value of β_1 is reduced continuously by 0.05 for each 6.9 MPa of strength in excess of 27.6 MPa (ACI Committee 318, 1983a).

If the tension steel yields ($f_s = f_y$) when the ultimate strength is attained, equilibrium of moments gives

$$M_u = 0.85abf'_c(d - 0.5a) + A'_s f'_s(d - d') \tag{3-1}$$

where M_u = ultimate moment
f'_c = specified compressive strength of concrete
f'_s = stress in compression steel
f_y = yield stress of tension steel
A'_s = area of compression steel

$$a = \text{depth of rectangular stress block}$$
$$b = \text{width of cross section}$$
$$d' \text{ and } d = \text{distances from extreme compression fiber to centroid of compression steel and to centroid of tension steel, respectively}$$

The equilibrium of forces acting on the section gives

$$a = \frac{A_s f_y - A'_s f'_s}{0.85 f'_c b} \tag{3-2}$$

where $A_s =$ the area of the tension steel.

In view of the strain distribution and $f-\varepsilon$ relationship, the stress f'_s is given as

$$f'_s = \varepsilon'_s E_s = 0.003 E_s \frac{a - \beta_1 d'}{a} \leqq f_y \tag{3-3}$$

where $E_s =$ Young's modulus of compression steel
$\beta_1 = a/c$
$c =$ distance from extreme compression fiber to neutral axis

Balanced flexural failure occurs when the tension steel reaches the yield stress f_y and the concrete reaches the extreme compression fiber strain of 0.003 simultaneously. At this instant, we have

$$a_b = \frac{0.003 \beta_1 d}{0.003 + f_y/E_s} \tag{3-4}$$

where $a_b =$ the depth of the rectangular stress block at the state of balanced flexural failure. In view of Eqs. (3-2) and (3-4), the balanced-reinforcement ratio ρ_b is obtained as follows:

$$\rho_b = \frac{0.85 f'_c \beta_1}{f_y} \left(\frac{0.003 E_s}{0.003 E_s + f_y} \right) + \frac{\rho' f'_s}{f_y} \tag{3-5}$$

where $\qquad \rho_b = \dfrac{A_s}{bd} \qquad \text{and} \qquad \rho' = \dfrac{A'_s}{bd} \tag{3-6}$

If the actual reinforcement ratio $\rho > \rho_b$, ultimate strength is reached when the concrete crushes; if $\rho < \rho_b$, the tension steel yields before the concrete crushes. As for T-section beams in most cases the neutral axis is located within the slab thickness so that the T section may be replaced by an equivalent rectangular section whose width is equal to the effective slab width, and the strength formulas described above are applicable.

3.3.3.3 Curvature Ductility of Beams   Available ductility is defined as the ratio of the curvature at the maximum carrying capacity ϕ_u to the curvature at the first yield of the tension steel ϕ_y (see Fig. 3-16). Substituting $\phi = \phi_y$ and $f_s = f_y$ in the strain and stress distributions in Fig. 3-16b leads to

$$\phi_y = \frac{f_y/E_s}{d(1 - k)} \tag{3-7}$$

where
$$k = \left[(\rho + \rho')^2 n^2 + 2\left(\rho + \rho'\,\frac{d'}{d}\right)n \right]^{\frac{1}{2}} - (\rho + \rho')n \tag{3-8}$$

and $n = E_s/E_c$ is the modular ratio.

In view of Fig. 3-16c, the ultimate curvature is

$$\phi_u = \frac{\varepsilon_c \beta_1}{a} \tag{3-9}$$

and thus

$$\frac{\phi_u}{\phi_y} = \frac{\varepsilon_c}{f_y/E_s}\,\frac{d(1 - k)}{a/\beta_1} \tag{3-10}$$

Since a value of $\varepsilon_c = 0.003$ is too conservative in ultimate-curvature calculations, it may be taken to be equal to 0.004. Equation (3-10) indicates that the ductility ϕ_u/ϕ_y increases with decreasing tension-steel content, increasing compression-steel content, decreasing steel yield stress, and increasing concrete strength (Park and Paulay, 1975). An effective way to increase ϕ_u/ϕ_y is to increase ε_c by the confining effect of transverse reinforcement. As to the relation between the confining-steel content and the value of ε_c, various empirical formulas have been proposed by several investigators (Park and Paulay, 1975).

3.3.3.4 Flexural Strength of Beam Columns

The ultimate strength P_0 of an axially loaded reinforced-concrete column is given as follows (ACI Committee 318, 1983):

$$P_0 = 0.85 f'_c(A_g - A_{st}) + A_{st}f_y \tag{3-11}$$

where A_g is the gross area of the section and A_{st} the total area of steel. Equation (3-11) assumes that ultimate strength is given as the strength of the concrete section plus the yield strength of the steel and that the concrete strength in the axially loaded column is equal to 0.85 times the compressive strength f'_c of a cylinder.

Ultimate strength of an eccentrically loaded column P_u is obtained according to ACI 318-83, from the same assumptions as in the beam-strength calculation. In view of Fig. 3-17, the equilibrium equations for the axial force and the bending moment become as follows:

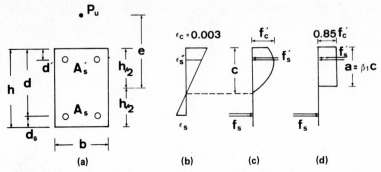

Figure 3-17 Eccentrically loaded column section at the ultimate state. *(a)* Cross section. *(b)* Strain. *(c)* Actual stresses. *(d)* Equivalent stresses.

$$P_u = 0.85f'_c ba + A'_s f'_s - A_s f_s \tag{3-12}$$

$$P_u e = 0.85f'_c ba(0.5h - 0.5a) + A'_s f'_s(0.5h - d') + A_s f_s(0.5h - d_s) \tag{3-13}$$

where h = the depth of the section, d_s = the distance from the extreme-tension fiber to the centroid of the tension steel, e = the load eccentricity, and f_s = the stress in the tension steel. f'_s is obtained from Eq. (3-3), and f_s is obtained similarly as follows:

$$f_s = \varepsilon_s E_s = 0.003E_s \frac{d\beta_1 - a}{a} \leqq f_y \tag{3-14}$$

Figure 3-18 illustrates an interaction between P_u and $P_u e$, calculated by Eqs. (3-12) and (3-13). These equations do not apply to the branch of

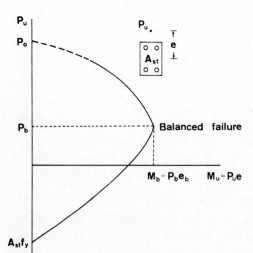

Figure 3-18 Interaction diagram for a beam column.

the interaction curve shown by a dashed line, in which the eccentricity is so small that the neutral axis lies outside the cross section. This portion must be constructed from other equations. P_b and $P_b e_b$ are calculated by substituting $f_s = f_y$ and $a = a_b$ into Eqs. (3-12) and (3-13), where a_b is given by Eq. (3-4). When the extreme fiber strain of the compression concrete becomes 0.003, the tension steel has already yielded if axial force is less than P_b but still remains elastic if axial force is greater than P_b.

3.3.3.5 Curvature Ductility of Beam Columns

The reduction of the ductility of reinforced-concrete beam columns due to the increase in axial force has been shown in Fig. 3-13. Figure 3-19 shows the interactions between N/bhf_c' and M/bh^2f_c' of the beam column, in which the values of N/bhf_c' are also plotted against arguments of $\phi_u h/\varepsilon_0$, $\phi_y h/\varepsilon_0$, and ϕ_u/ϕ_y, where ϕ_u and ϕ_y are, respectively, the curvatures at maximum carrying capacity and at the first yield of the tension steel and $\varepsilon_b = 0.002$. Similar diagrams can be found in Pfrang, Siess, Sozen (1964). It is observed that ductility is reduced as the increasing axial force approaches the balanced-failure state at which ϕ_u/ϕ_y becomes unity. Therefore, transverse confining reinforcement is needed to ensure sufficient ductility of the beam column under relatively large axial force, and a method has been proposed for calculation of the confining-reinforcement content needed for the required value of ductility (Park and Paulay, 1975). ACI 318-83 recommends that the concrete core of a beam column be confined by special transverse reinforcement when axial force is greater than $A_g f_c'/10$ (ACI Committee 318, 1983), where A_g is the gross area of the section.

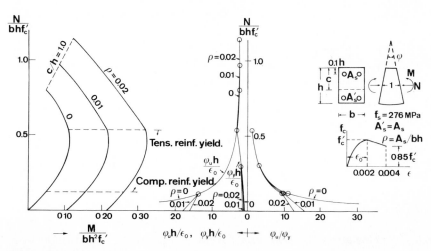

Figure 3-19   Strength and ductility of an eccentrically loaded column section.

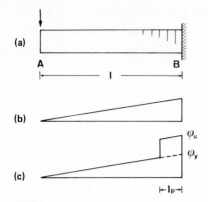

Figure 3-20 Member with ultimate curvature reached. *(a)* Member. *(b)* Moment diagram. *(c)* Curvature diagram. [*From R. Park and T. Paulay, Concrete structures, in E. Rosenblueth (ed.),* Design of Earthquake Resistant Structures, *Pentech Press, London, 1980, pp. 142–194.*]

3.3.3.6 Deflection Ductility of Members   At the ultimate-failure state of a cantilever subjected to a transverse load at its tip as shown in Fig. 3-20, the plastic zone develops near the fixed end where the bending moment exceeds the yield moment, and excessive curvature is concentrated in this zone. The deflection at the tip is obtained as the first moment of the area of the curvature diagram at about point A. Let us suppose that the inelastic curvature distribution is idealized as shown in Fig. 3-20c; the ratio μ of the ultimate deformation Δ_u to the deflection at the first yield Δ_y is given as follows (Park and Paulay, 1980):

$$\mu = \frac{\Delta_u}{\Delta_y} = 1 + \left(\frac{\phi_u}{\phi_y} - 1\right) \frac{3\ell_p(\ell - 0.5\ell_p)}{\ell^2} \tag{3-15}$$

where ℓ = the length of the cantilever and ℓ_p = the length of the plastic zone. Equation (3-15) indicates that the member ductility μ decreases with decreasing ℓ_p/ℓ for a given value of the curvature ductility. In other words, greater curvature ductility is needed to ensure adequate member ductility μ as ℓ_p/ℓ becomes smaller. An empirical formula for the calculation of ℓ_p is proposed as follows (Corley, 1966; Mattock, 1967):

$$\ell_p = 0.5d + 0.05Z \tag{3-16}$$

where d = the effective depth of member and Z = the distance from the critical section to the point of contraflexure. For example, if ℓ/d is assumed to be equal to 5 in Fig. 3-20, ℓ_p/ℓ becomes 0.15 from Eq. (3-16). Then if it is required that $\Delta_u/\Delta_y = 4$, substituting $\Delta_u/\Delta_y = 4$ and $\ell_p/\ell = 0.15$ into Eq. (3-15) leads to $\phi_u/\phi_y = 8.2$. The required value for curvature ductility is thus relatively large.

3.3.3.7 Calculation of Hysteresis Curves   Theoretical load-deflection curves for cyclically loaded members as shown in Fig. 3-14 can be accurately obtained by the step-by-step method based on moment-curvature relationships for the cross sections. It is convenient to divide a member into a number of longitudinal elements and to assume the moment to be uniform over the length of each element. Let us consider a cantilever subjected to constant axial force and to cyclic transverse load at the tip as an example for describing the calculation procedure. First, a trial value of bending moment in the first element at the fixed end is set, and then the curvature in this element is calculated by the procedure described later. By integrating the curvature over the length of the longitudinal element with boundary conditions at the fixed end, the rotation angle and the deflection at the end of the first element are calculated, and consequently the moment for the next element is determined from the equilibrium of the first element including the secondary-moment (P-Δ-moment) effect. Repetition of this procedure to the free end leads to a deflected configuration with a bending-moment distribution which may not satisfy the zero-moment condition at the free end. Thus, iterative calculations adjusting the trial value of the fixed-end moment are needed until the zero-moment condition is satisfied at the free end. Once this has been satisfied, the transverse load is calculated from the overall equilibrium of the cantilever. It should be noted that this method of determining the load-deflection relation contains a certain error, since the effect of shear deformation is not taken into account.

The procedure for calculating deflections explained above involves calculation of the moment-curvature relation. The most convenient method of determining the internal forces is to sum the stresses acting on a number of horizontal elements into which the section is divided as shown in Fig. 3-21. The procedure for determining the curvature for a given level of the external moment is as follows. Once trial values for the extreme fiber strain of the compression concrete ε_{cm} and the neutral axis ratio k have been given, the curvature ϕ and the strain in each element ε_i are determined as shown in Fig. 3-21. The stress in each element f_i can be determined from the prescribed models of the stress-strain relationships, such as are shown in Figs. 3-3 and 3-6, and consequently the internal axial force and bending moment are calculated by summing the

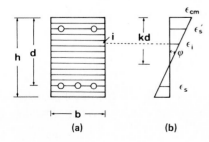

Figure 3-21 Discrete elements for a cross section. *(a)* Cross section. *(b)* Strain distribution.

stresses acting on discrete elements. If these are not equal to the external axial force and bending moment, respectively, the trial values for ε_{cm} and k must be adjusted until equilibrium between internal and external forces is satisfied.

3.3.4 Shear Behavior of Members

3.3.4.1 Load-Deformation Relationships   Figure 3-22 shows the relationship between the shear force V and the chord rotation R for reinforced-concrete members subjected to the axial force N and the antisymmetrical bending moment M with shear, where axial force is kept constant and equal to 0 or 30 percent of the ultimate strength of a centrally loaded column (Wakabayashi, Minami, et al., 1974). In the figure, s designates the spacing of shear reinforcement, A_v the area of shear reinforcement within a distance s, and b and h the width and depth of the section, respectively. It is observed that ductility increases with increasing value of A_v/bs, and shear failure becomes more brittle as axial force becomes larger.

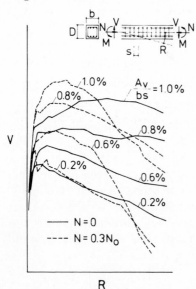

Figure 3-22 Load-deformation relationships for beam columns failing in shear. [*From M. Waka-bayashi, K. Minami, et al., Some tests on elastic-plastic behavior of reinforced concrete frames with emphasis on shear failure of columns,* **Annuals: Disaster Prevention Res. Inst.,** *Kyoto University, no. 17-B, 171–189 (1974; in Japanese).*]

The behavior of members failing in shear when subjected to constant axial force and repeated bending with shear is indicated in Fig. 3-23 (Wakabayashi and Minami, 1976). Comparison with the flexural behavior of members shown in Fig. 3-14 highlights the following characteristics of members failing in shear:

1. Strength deterioration after maximum carrying capacity is drastic, and ductility is small.

2. Hysteresis loops show a pinched shape, and energy-dissipation capacity is small.

3. Strength degradation due to load repetition is severe.

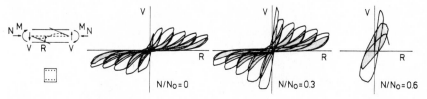

Figure 3-23 Relationships between shear force and rotation angle for beam columns failing in shear under constant axial force and alternately repeated antisymmetrical bending moment. [*From M. Wakabayashi and K. Minami, Experimental studies on hysteretic characteristics of steel reinforced concrete columns and frames, Proc. Int. Symp. Earthquake Eng., University of Missouri–Rolla, St. Louis, 1,467–480 (1976).*]

3.3.4.2 Diagonally Reinforced Concrete Members   Deep beams, such as coupling beams for shear walls, often fail in shear, and their behavior is not very much improved even if stirrups are increased. Research in New Zealand has provided a method for improving the ductility of such deep beams by diagonal reinforcement, and it has already been applied in practice (Park and Paulay, 1975). The author has also used diagonal reinforcement in beam columns and tested them (Minami and Wakabayashi, 1980; Wakabayashi and Minami, 1980a). Figure 3-24 compares the hysteretic behavior of beam columns with parallel and diagonal reinforcement. Reinforcement content is kept equal for both members. The diagonally reinforced beam column shows very advantageous characteristics:

1. Large shear strength
2. Large ductility
3. Large energy-dissipation capacity
4. Gradual degradation
5. Failure by flexure

In design practice, combined use of parallel and diagonal reinforcement may be realistic. Monotonic load-deformation curves are shown in Fig. 3-25 for various values of the ratio of diagonal-reinforcment content to total-reinforcement content. Diagonal reinforcement clearly improves ductility as well as shear strength (Wakabayashi, Minami, et al., 1981). A similar trend is to be seen in members under alternately repeated loading.

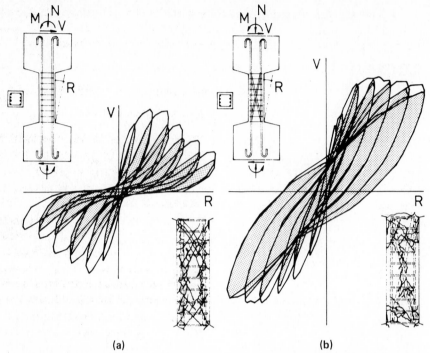

(a)   (b)

Figure 3-24 Relationships between moment and rotation angle for beam columns with parallel or diagonal reinforcement under constant axial force and alternately repeated bending moment with shear. *(a)* Column with parallel reinforcement. *(b)* Column with diagonal reinforcement. [*From M. Wakabayashi and K. Minami, Seismic resistance of diagonally reinforced concrete columns,* Proc. Seventh World Conf. Earthquake Eng., *Istanbul, 6, 215–222 (1980).*]

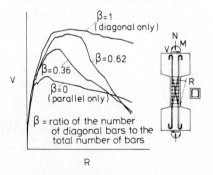

Figure 3-25 Effect of the amount of diagonal reinforcement on load-deflection relationships for beam columns. [*From M. Waka-bayashi, K. Minami, et al., Effectiveness of diagonal reinforcement applied to reinforced concrete columns subjected to shear force,* Proc. Third Ann. Conv. Japan Concr. Inst., *445–448 (1981; in Japanese).*]

3.3.4.3 Shear Strength   The shear resistance of reinforced-concrete members is mainly provided by two mechanisms, a truss action and an arch action, as shown in Fig. 3-26 (Bresler and MacGregor, 1967).

The truss mechanism for resisting flexure and shear is an old concept. The concrete compression zone and the tension reinforcement form the

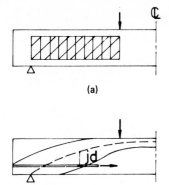

(a)

(b)

Figure 3-26 Shear-resistance mechanisms. *(a)* Truss mechanism. *(b)* Arch mechanism.

top and bottom chords of the analogous truss, and the web consists of stirrups acting as tension vertical members and concrete struts acting as compression diagonal members, as shown in Fig. 3-26a. The compression force in a diagonal concrete strut is brought into equilibrium with the tension force in the bottom chord member by a bond, and shear force is carried by the stirrups only. The shear strength of the member is thus the smallest value as limited by the following: stirrup strength, bond strength, and compression strength of the concrete strut.

In the arch-mechanism concept bond action is ignored, and the position of the line of thrust is considered to vary along the length of the member, as shown in Fig. 3-26b. The shear is resisted by arching action in the concrete, and the ultimate-failure state is reached when the concrete crushes.

The failure mechanism of a reinforced-concrete member subjected to axial force, bending moment, and shear is complex, and a widely accepted strength formula has not yet been developed. However, the ultimate strength of a member subjected to combined loads may be estimated with reasonable accuracy by a theoretical approach, in which the ultimate strength of a member as shown in Fig. 3-27a is simply assumed to be the sum of the strength provided by the truss mechanism in Fig. 3-27b and the arch mechanism in Fig. 3-27c (Minami and Wakabayashi, 1981). The width of the concrete section for each mechanism is determined so as to obtain maximum ultimate strength of the member.

The approach described above, when applied to a member of a given length, produces the results shown in Fig. 3-28 in the form of a strength-interaction diagram with axial force N plotted against shear force V. The solid and dashed lines indicate the strengths of the member with and without hoops, respectively. The dash-and-dot line is obtained from the strength-interaction curve of moment versus axial force for the cross

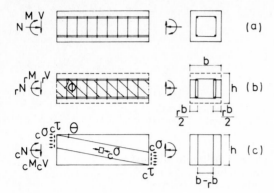

Figure 3-27 Resistance mechanisms in a reinforced-concrete member. *(a)* Member. *(b)* Truss mechanism. *(c)* Arch mechanism. [*From K. Minami and M. Waka-bayashi, Rational analysis of shear in reinforced concrete columns, IABSE Colloquium, Delft, 1981,* Advanced Mechanics of Reinforced Concrete: Final Report, *International Association of Bridge and Structural Engineering, Delft, Netherlands, 603−614 (1981).*]

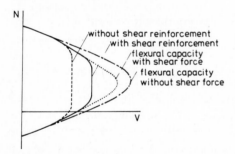

Figure 3-28 Shear strength of reinforced-concrete beam columns. [*From K. Minami and M. Wakabayashi, Rational analysis of shear in reinforced concrete columns, IABSE Colloquium, Delft, 1981,* Advanced Mechanics of Reinforced Concrete: Final Report, *International Association of Bridge and Structural Engineering, Delft, Netherlands, 603−614 (1981).*]

section in which the shear effect is neglected. The dotted line indicates the strength of a member in which the hoop content is sufficiently large to ensure that flexural failure with yielding of longitudinal reinforcement takes place prior to shear failure. It is noticed that the shear effect reduces strength even though the hoop content is sufficient to make the member fail in flexure and that shear strength is affected by axial-force intensity.

The shear strength of a member in which diagonal and parallel reinforcements are used together, as shown in Fig. 3-25, is calculated as the sum of the strength of a member with parallel reinforcement and the strength of a truss containing the diagonal reinforcement. Figure 3-29 indicates the interaction curves for such members, together with experimental data. The theoretically obtained strengths agree well with experimental results. Note that the increase in strength due to diagonal reinforcement is more apparent in beam columns than in beams.

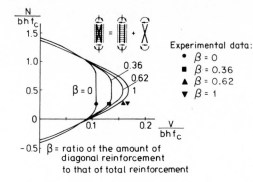

Experimental data:
- $\beta = 0$
- $\blacksquare$ $\beta = 0.36$
- $\blacktriangle$ $\beta = 0.62$
- $\blacktriangledown$ $\beta = 1$

β = ratio of the amount of diagonal reinforcement to that of total reinforcement

Figure 3-29 Shear strength of diagonally reinforced beam columns. [*From M. Wakabayashi, K. Minami, et al., Effectiveness of diagonal reinforcement applied to reinforced concrete columns subjected to shear force*, Proc. Third Ann. Conv. Japan Concr. Inst., 445–448 (1981; in Japanese).]

For the purpose of proportioning the cross section, empirical strength formulas are often used instead of a theoretical approach such as the one explained above. The ACI Code recommends an empirical formula which gives shear strength as the sum of the strength provided by concrete at the onset of cracking V_c plus the strength provided by shear reinforcement V_s (ACI Committee 318, 1983a), that is,

$$V_u = V_c + V_s \qquad (3\text{-}17)$$

The shear strength of concrete is written as

$$V_c = v_c b_w d \qquad (3\text{-}18)$$

where v_c = the nominal shear stress of concrete at cracking, b_w = the web width, and d = the effective depth of the member. The value of v_c varies according to the ratio of the shear force to the bending moment, and the ACI Code recommends various empirical formulas for v_c expressed in terms of this ratio, together with the simpler formulas given below:

$$v_c = 0.17\sqrt{f_c'} \qquad \text{MPa} \qquad (3\text{-}19)$$

for members without axial force,

$$v_c = 0.17\left(1 + 0.073\,\frac{N_u}{A_g}\right)\sqrt{f_c'} \qquad \text{MPa} \qquad (3\text{-}20)$$

for members subjected to compression, and

$$v_c = 0.17\left(1 + 0.29\,\frac{N_u}{A_g}\right)\sqrt{f_c'} \qquad \text{MPa} \qquad (3\text{-}21)$$

for members subjected to tension.

In these expressions N_u = the axial force and is taken to be positive for compression, A_g = the gross area of the section, and f_c' is expressed in megapascals. The shear strength provided by shear reinforcement in the form of stirrups or hoop ties is given as

$$V_s = A_v f_y d/s \qquad (3\text{-}22)$$

where A_v = the area of shear reinforcement within a distance s, s = the shear reinforcement spacing, and f_y = the yield stress of steel reinforcement. Then Eq. (3-17) becomes

$$V_u = v_c b_w d + A_v f_y d/s \qquad (3\text{-}23)$$

The ACI specifies the maximum value for V_s as $0.66\sqrt{f_c'}\, b_w d$ (MPa), since the value of V_u obtained from Eq. (3-23) otherwise becomes excessive if A_v is very large (ACI Committee 318, 1983a).

For a member with diagonal and parallel reinforcement,

$$V_u = v_c b_w d + A_v f_y d/s + 2A_{sd} f_y \sin\theta \qquad (3\text{-}24)$$

where A_{sd} = the area of diagonal reinforcement in tension, which is assumed to be equal to that in compression; and θ = the angle between diagonal reinforcement and the longitudinal axis of the member. Note that Eq. (3-24) is obtained by adding the shear strength of an analogous truss consisting of diagonal reinforcement, $2A_{sd} f_y \sin\theta$, to the shear strength of a member with parallel reinforcement given by Eq. (3-23).

3.3.5 Shear Walls

3.3.5.1 Squat Shear Walls When shear is applied to a wall which has either no frame at all or a weak surrounding frame, as shown in Fig. 3-30a, the load-deformation relationship for the wall is as indicated in Fig. 3-30b. The capacity of a wall which has inadequate reinforcement or no reinforcement at all is reduced after cracking occurs, and the curve extends down to point A. On the other hand, if the reinforcement content is adequate, the curve extends upward to point B because tension reinforcement tends to prevent free rotation of the concrete diagonal struts and expansion of the wall, as shown in Fig. 3-30a.

The shear strength of a squat shear wall at diagonal cracking V_c is given as follows (Tall Building Committee 21D, 1979):

$$V_c = 0.1t\ell f_c' \qquad (3\text{-}25)$$

where t and ℓ = the thickness and horizontal length of the wall, respectively; and f_c' = the compressive strength of concrete. The shear strength of the wall at the ultimate state V_w is provided only by wall reinforcement and is given by

$$V_w = \rho_w t\ell f_y \leqq 0.18t\ell f_c' \qquad (3\text{-}26)$$

where ρ_w = the horizontal-reinforcement-content ratio, which is assumed to be equal to the vertical-reinforcement-content ratio; and f_y = the

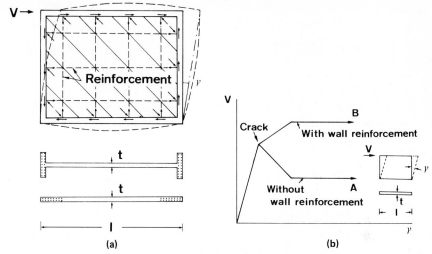

Figure 3-30 Diagonal cracks and load-deformation relationships for a shear wall. *(a)* Diagonal cracks and the effect of wall reinforcement. *(b)* Idealized load-deformation relationships for a shear wall.

yield stress of the wall reinforcement. The term $0.18t\ell f'_c$, which is empirically derived, indicates that the shear strength of a shear wall is controlled by the crushing strength of the concrete diagonal strut if the wall-reinforcement content is large.

If the surrounding frame is strong enought to carry the forces delivered from the concrete diagonal struts, the infilled wall is able to resist the shear load until sliding failure takes place with crushing of the concrete strut as shown in Fig. 3-31a, even though no wall reinforcement is provided. In this case, ultimate shear strength of the wall is thus given by

$$V_u = 0.18t\ell f'_c \tag{3-27}$$

However, the surrounding frame is not usually this strong, and wall resistance then depends on the wall-reinforcement content. The ultimate shear strength of the infilled shear wall V_u is given as the sum of the strength of the wall plus the strength of the columns:

$$V_u = V_w + \Sigma V_{c\ell} \tag{3-28}$$

where V_w is given by Eq. (3-26) and V_{cl} = the shear strength of a single column. Figure 3-31 shows the failure mode and the hysteretic behavior of such an infilled wall. It is observed that stiffness is gradually reduced after the onset of cracking, and maximum strength is attained when the story drift angle is equal to about 0.004 rad. Various attempts have been

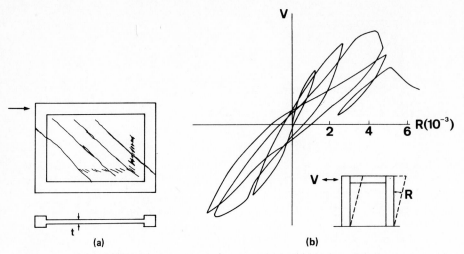

Figure 3-31 Load-deformation relationship for an infilled wall. *(a)* Failure mode. *(b)* Hysteresis curve.

made to improve brittle behavior and increase the ductility of infilled walls (see Sec. 3.5.5).

3.3.5.2 Cantilever Shear Walls   A cantilever shear wall behaves similarly to a single reinforced-concrete beam column and shows various failure modes as illustrated in Fig. 3-32 (Park and Paulay, 1980).

Failure Due to Overturning Moment. The flexural strength M_u of a rectangular shear wall subjected to axial force as shown in Fig. 3-32a is given approximately as follows (Cardenas, Hanson, et al., 1973):

$$M_u = 0.5A_s f_y \ell_w \left(1 + \frac{N_u}{A_s f_y}\right)\left(1 - \frac{c}{\ell_w}\right) \tag{3-29}$$

where
$$\frac{c}{\ell_w} = \frac{q + \alpha}{2q + 0.85\beta_1}$$

$$q = \frac{A_s f_y}{\ell_w h f_c'}$$

$$\alpha = \frac{N_u}{\ell_w h f_c'}$$

A_s = area of total vertical wall reinforcement
f_y = yield stress of wall reinforcement
N_u = compressive axial force

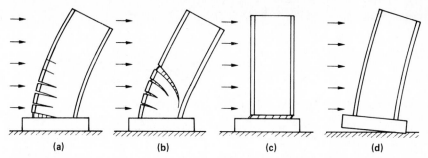

Figure 3-32 Failure modes for a cantilever shear wall. *(a)* Flexural failure. *(b)* Shear failure. *(c)* Sliding failure. *(d)* Rotation of foundation. [*From R. Park and T. Paulay, Concrete structures, in E. Rosenblueth (ed.),* Design of Earthquake Resistant Structures, *Pentech Press, London, 1980, pp. 142–194.*]

ℓ_w = horizontal length of wall

c = distance from extreme compression fiber to neutral axis

h = thickness of shear wall

f_c = specified compressive strength of concrete

β_1 = 0.85 for strength f_c up to 27.6 MPa and is reduced continuously to a rate of 0.05 for each 6.89 MPa of strength in excess of 27.6 MPa

The axial force in the wall due to the vertical load is usually smaller than the value at the balanced-failure condition. The tension steel therefore yields in the ultimate-failure state, and large ductility and energy-dissipation capacity can be expected as shown in Fig. 3-33 (Fiorato, Oesterle, and Carpenter, 1976). However, if axial force becomes large, the neutral axis approaches the tension side (Fig. 3-34), causing a large extreme fiber strain in the compression concrete and reduced ductility. Ductility is ensured by confining the concrete in the compression zone at the base of the wall. The situation is smaller for flanged shear walls.

Shear Failure. A shear wall with a small aspect ratio is likely to fail in shear, with diagonal cracking as shown in Fig. 3-32*b*. Diagonal tension failure occurs if horizontal reinforcement content is small, while diagonal compression failure occurs if reinforcement is adequate. Shear strength is provided by the concrete and the horizontal reinforcement, as in Eq. (3-17) for the beam column. However, shear resistance provided by the concrete in the vicinity of a plastic hinge which forms due to flexure deteriorates with load repetition, and thus concrete shear resistance is best ignored (Park and Paulay, 1980).

The shear strength of the wall increases as the aspect ratio becomes smaller. This characteristic is allowed for in the equations of the ACI Code. However, the code also permits this effect to be ignored. Shear-

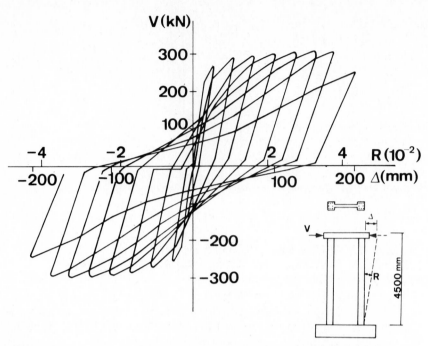

Figure 3-33 Load-deflection relationship for a cantilever shear wall failing in flexure. [*From A. E. Fiorato, R. G. Oesterle, Jr., and J. E. Carpenter, Reversing load tests of five isolated structural walls,* Proc. Int. Symp. Earthquake Struct. Eng., *University of Missouri–Rolla, St. Louis, 437–453 (1976).*]

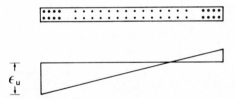

Figure 3-34 Strain distribution in a shear wall.

strength formulas for the beam column, that is, Eqs. (3-17) through (3-21), can also be used to calculate the shear strength of the wall, neglecting the compressive force N_u. The effective wall length d should be taken to be equal to 0.8 times the horizontal length ℓ_w. The total nominal shear strength of concrete must not be taken to be greater than $0.83\sqrt{f_c'}$ (MPa), since failure occurs because of concrete crushing if shear reinforcement content is large.

The Appendix of ACI 318-83 gives the same equation derived from Eqs. (3-19) and (3-23) for the shear strength of walls.

Sliding-Shear Failure. In the sliding-shear failure mode shown in Fig. 3-32c, the shear wall moves horizontally. To prevent this type of failure, vertical reinforcement uniformly spaced in the wall, as well as diagonal reinforcement, is effective (see Sec. 4.6.2.7). Sliding-shear failure also occurs at construction joints, for which vertical reinforcement is again effective.

Uplifting of the Foundation. The hysteretic load-deflection relationship for a multistory shear wall in which uplifting of the foundation occurs as shown in Fig. 3-32d is shown in Fig. 3-35. It is observed that loading and unloading curves follow almost the same path, and hence energy-dissipation capacity is extremely small (Wakabayashi, Minami, et al., 1978).

3.3.5.3 Shear Walls with Openings   Shear failure often occurs in the connecting beams of a shear wall with openings, as shown in Fig. 3-36a. Although energy-dissipation capacity is improved by the presence of a large quantity of stirrups, high ductility cannot be expected. In such a case diagonal reinforcement is much more effective (Park and Paulay, 1975; Paulay, 1972, 1980). The same may be said for shear failure which occurs in wall columns as shown in Fig. 3-36b.

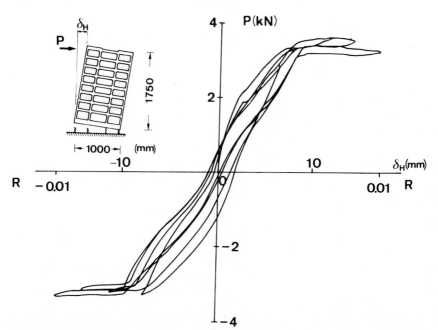

Figure 3-35 Load-deflection relationship for a frame with uplifting of foundations. [*From M. Wakabayashi, K. Minami, et al., Experimental study on the elasto-plastic behavior of reinforced concrete frames with multi-story shear walls, Proc. Arch Inst. Japan Ann. Conv., 1467−1460 (1978; in Japanese).*]

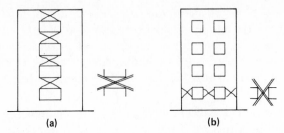

Figure 3-36 Shear failure of walls with openings and diagonal reinforcement. *(a)* Connecting beams. *(b)* Wall columns.

3.3.6 Connections

When a horizontal load is applied to a reinforced-concrete frame, a typical beam-to-column connection is subjected to antisymmetrical moments as shown in Fig. 3-37*a*. Shear stress is induced in the joint block, and diagonal cracks occur at a certain load level. After cracking, the core concrete acts as a diagonal compression strut as shown in Fig. 3-37*b* and the load increases further. Finally, the concrete block crushes if the flexural strengths of adjacent columns and beams are sufficiently large (Corley and Hanson, 1969; Fenwik and Irvine, 1977; Uzumeri, 1977; Paulay, Park, and Priestley, 1978; Gavrilovic, Velkov, et al., 1980; Paulay and Scarpas, 1981; Jirsa, 1981; Jirsa, Meinheit, and Woolen, 1975). Figure 3-38 shows a sample of hysteretic behavior for such a beam-to-column connection (Wakabayashi, Nakamura, and Matsuda, 1977). Characteristics similar to those of a reinforced-concrete member failing in shear are observed: low ductility, low energy-dissipation capacity, severe strength degradation, etc.

The bending moments and the shear forces acting around the joint block can be separated into antisymmetrical forces due to the horizontal load and symmetrical forces due to the vertical load. The shear in the

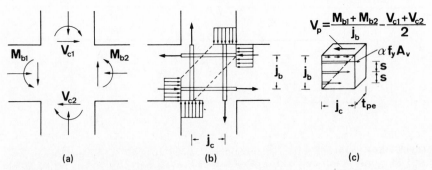

Figure 3-37 Resisting mechanism in a connection panel. *(a)* Forces. *(b)* Stresses on the panel. *(c)* Resistance of the panel.

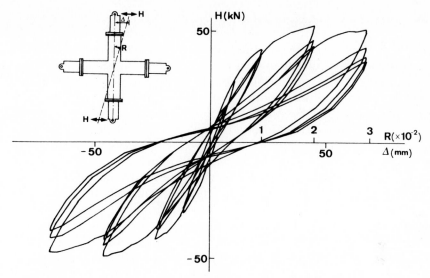

Figure 3-38 Load-deformation relationship for a beam-to-column connection. [*From M. Wakabayashi, T. Nakamura, and H. Matsuda, Experimental study on the stress transmission and load carrying capacity of reinforced concrete beam-to-column connections*, Proc. Arch. Inst. Japan Ann. Conv., *1781–1782 (1977; in Japanese).*]

block is related to antisymmetrical forces. In view of Fig. 3-37c, the shear force acting on the top boundary of the block is given as

$$V_p = \frac{M_{b1} + M_{b2}}{j_b} - \frac{V_{c1} + V_{c2}}{2} \tag{3-30}$$

where j_b = the internal lever arm of the beam, M_{b1} and M_{b2} = the bending moments at beam ends, and V_{c1} and V_{c2} = the shear forces at column ends. Empirical formulas proposed for the shear strength of the block are (Kamimura, 1975)

$$V_p = v_c t_{pe} j_c + 0.5 f_y A_v j_c / s \tag{3-31}$$

$$\text{and} \quad v_c = \begin{cases} (0.78 - 0.016 f_c') f_c' & \text{for} \quad f_c' \leqq 23.9 \quad \text{MPa} \\ 9.3 & f_c' > 23.9 \quad \text{MPa} \end{cases} \tag{3-32}$$

where j_c = the internal level arm of the column, t_{pe} = the effective thickness of the block (taken to be equal to the average of beam and column widths), and A_v = the area of hoops at spacing s. The second term on the right-hand side of Eq. (3-31) implies that the stress in a hoop at the ultimate state is equal to half of the yield stress of the steel. Equations given above are for a cross-shaped beam-to-column connection. The value of v_c for T-shaped or L-shaped connections is less than the value given by Eq. (3-32) since fewer members are available to provide reactions to the compressive force sustained by the diagonal concrete strut.

A longitudinal reinforcing bar extending through the joint block as shown in Fig. 3-39a is pushed at one end of the panel and pulled at the other. Bond stress acting around the bar resists the slip. If bond strength is not sufficient, slip takes place and reduces compressive stress in the bar so that tensile stress is sometimes developed on the bar's compression side, as shown in Fig. 3-39a. Figure 3-39b shows an M-N interaction curve for the section in which slip of the longitudinal reinforcement is taken into account. It is observed that the strength of a beam column is reduced by slip while that of a beam is not much affected (Wakabayashi, Nakamura, and Matsuda, 1977). It has been known that slip increases deformation and decreases energy-dissipation capacity (Viwathanatepa, Popov, and Bertero, 1979).

The strength of a beam-to-column connection also depends on anchorage details. Since bottom reinforcement in a beam is subjected to tension under earthquake loading, sufficient anchorage length must be provided by making the length ℓ nearly equal to column depth h as shown in Fig. 3-40. Otherwise flexural strength at the beam end is reduced for the case of bending moment, causing tension in bottom reinforcement. If sufficient anchorage length cannot be provided because the depth h is too small, anchorage in a stub beam as indicated by the dashed lines in Fig. 3-40 is effective (Park and Paulay, 1975).

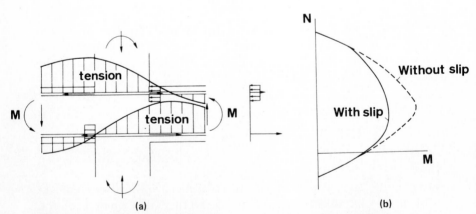

(a)   (b)

Figure 3-39 Slip of the main reinforcement in a connection panel. (a) Stress distribution in reinforcement. (b) Interaction curve for a member. [*From M. Wakabayashi, T. Nakamura, and H. Matsuda, Experimental study on the stress transmission and load carrying capacity of reinforced concrete beam-to-column connections, Proc. Arch. Inst. Japan Ann. Conv., 1781–1782 (1977; in Japanese).*]

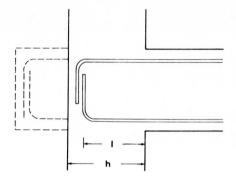

Figure 3-40 Anchorage of the main reinforcement into the joint core.

3.3.7 Systems

Once the behavior of structural components and joints, such as beams, beam columns, and connections, is known, the behavior of the system as a whole can be obtained by integration.

The hysteretic load-deflection curves for full-scale frames tested under horizontal loading are shown in Fig. 3-41. While the columns fail in flexure in case *a*, the short columns fail in shear in case *b*, in which the frame involves long and short columns together (Aoyama, Sugano, and Nakata, 1970).

A procedure for determining the hysteretic curve of a frame is as follows:

1. An idealized trilinear relation between moment and rotation, as shown in Fig. 3-42, is assumed for each member.

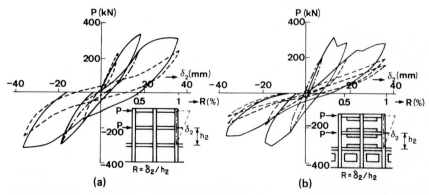

Figure 3-41 Load-deflection relationships for frames. (*a*) Frame failing in flexure. (*b*) Frame failing in shear. [*From H. Aoyama, S. Sugano, and S. Nakata, Vibration and static loading tests of Hachinohe Technical College, part 2: Static loading tests,* Trans. Arch. Inst. Japan, *no. 169, 33−41 (1970; in Japanese).*]

2. The ultimate strength of each column is calculated for the appropriate type of failure mechanism, i.e., a column failure or a beam failure at each joint, as shown in Fig. 3-43, and for each story the shear at the ultimate state is obtained by summing the shear forces carried by all columns at that story.

3. The inelastic stiffness of the overall system is calculated from inelastic stiffnesses of individual components.

4. A skeleton curve of the system, similar to the one shown in Fig. 3-42, is then developed, and hysteretic curves are obtained from this skeleton curve by the rules given in Sec. 2.5.1.

In this method of analysis the load-deflection curve is calculated from a failure mechanism which is locally determined at the joints. This method gives a good approximation of the behavior of a system in which the column-failure mechanism actually occurs, but the behavior of a system with the beam-failure mechanism cannot be well estimated. An improved method of calculating system stiffness directly from skeleton curves of components has been proposed (Umemura, Aoyama, and Takizawa, 1974). However, no method is yet available for a system in which the strength of the components either deteriorates after maximum-load-carrying capacity has been attained or degrades owing to load repetition, as, for example, in members failing in shear.

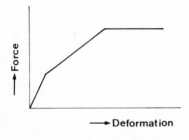

Figure 3-42   Force-deformation relationship.

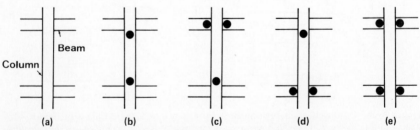

Figure 3-43   Failure mechanism. *(a)* Column-and-beam assemblage. *(b)* Column ends yield. *(c)* Column bottom yields. *(d)* Column top yields. *(e)* Beam ends yield.

3.3.8 Behavior of Prestressed-Concrete Structures

3.3.8.1 Hysteretic Behavior   Provided compressive-concrete strain has not become too large during the loading process, the cracks developed in a prestressed-concrete member close again and deformations return almost to zero when the load is removed. The hysteretic load-deformation curve becomes S-shaped, and energy-dissipation capacity is small, as shown in Fig. 3-44 (Muguruma, Watanabe, and Nagai, 1978). Therefore, the response to earthquake loading is severe. For example, it has been reported that the dynamic response of a prestressed-concrete frame to the north-south component of the El Centro earthquake (1940) is as much as 130 percent of the response of an ordinary reinforced-concrete frame designed for the same conditions (Park and Paulay, 1980). From this point of view, prestressed-concrete structures must be designed for magnified loads, and sufficient deformation capacity must be provided even for nonstructural members to allow them to follow large deformations.

The use of ordinary reinforcing bars together with prestressing steel makes structural behavior similar to that of an ordinary reinforced-concrete structure, and earthquake-resistant capacity is improved. The use of partially prestressed members is also effective (Park, 1980). A frame consisting of prestressed-concrete beams and reinforced-concrete columns behaves more like a reinforced-concrete frame than a frame consisting only of prestressed-concrete members, and mixed use of prestressed concrete and reinforced concrete therefore improves earth-

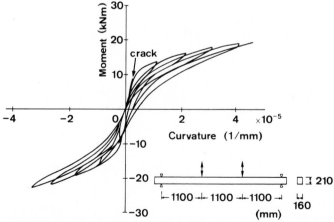

Figure 3-44 Moment-curvature relationship for a prestressed-concrete beam. [*From H. Muguruma, F. Watanabe, and E. Nagai, A study on the hysteretic characteristics of prestressed concrete statically indeterminate rigid frame under repeated horizontal load, Rep. Kinki Branch Arch. Inst. Japan, 73−76 (1978; in Japanese).*]

quake-resistant capacity. Stress in an unbonded prestressing cable is uniformly distributed along its length; hence the cable dissipates almost no energy, and this is undesirable for earthquake resistance.

3.3.8.2 Strength   The ultimate moment M_u of a prestressed-concrete member with bonded prestressing steel and with either a rectangular section or an I-shaped section with the neutral axis lying within the flange, is given by

$$M_u = A_{ps} f_{ps} \left(d - \frac{a}{2} \right) \tag{3-33}$$

and
$$a = \frac{A_{ps} f_{ps}}{0.85 f'_c b} \tag{3-34}$$

where A_{ps} = the area of prestressing steel, d = the distance from the extreme-compression fiber to the centroid of prestressing steel, b = the width of the compression face of the member, and f_{ps} = the stress in prestressing steel at the ultimate state. When effective steel prestressing is not less than 0.5 times the tensile strength of steel f_{pu}, the value of f_{ps} may be calculated (ACI Committee 318, 1983) as

$$f_{ps} = f_{pu}(1 - \xi) \tag{3-35}$$

where ξ = function of A_{ps} and area of nonprestressed reinforcement.

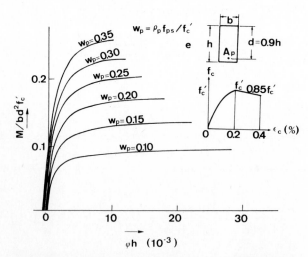

Figure 3-45 Moment-curvature relationships for prestressed-concrete beams with various amounts of prestressing bars. [*From S. Inomata, Background of the FIP specification for aseismic design of prestressed concrete structures, Prestressed Concr., 21(4), 62–71 (1979; in Japanese).*]

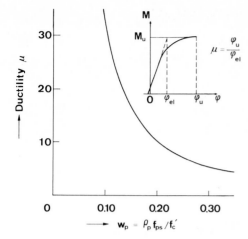

Figure 3-46 Relationship between w_p and ductility. [*From S. Inomata, Background of the FIP specification for aseismic design of prestressed concrete structures,* Prestressed Concr., **21**(4), 62–71 (1979; in Japanese).]

3.3.8.3 Ductility The ductility of prestressed-concrete members varies with the amount and position of the prestressing steel. For singly-prestressed-concrete members as shown in Fig. 3-45, moment strength increases with increasing value w_p ($= \rho_p f_{ps}/f'_c$); however, ductility decreases since the distance from the extreme-compression fiber to the neutral axis also increases (Inomata, 1979; Park and Thompson, 1977). Figure 3-46 shows a relation between w_p and the ductility factor μ, which indicates that μ can be expected to be at least 10 if the value of w_p is limited to less than 0.2 in the design.

Ductility can usually be improved by adding reinforcing steel to one or both faces of a prestressed section. The ductility of a doubly-prestressed-concrete member is not affected by the value of w_p.

3.3.9 Earthquake Damage

The main causes of earthquake damage have been described and listed in Sec. 3.3.1. Several examples of such damage are given in the following figures:

Figure 3-47. An example of factor 3 in the list in Sec. 3.3.1. Shear failure is in a column which is shortened by interaction with the spandrel walls. This is one of the most frequent failure modes in reinforced-concrete structures.

Figure 3-48. Brittle failure in a corner column which was transversely reinforced by tied hoops. Other columns displayed ductile behavior because they contained spiral reinforcement (Frazier, Wood, and Housner, 1971). This structure is also an example of factor 7 in Sec. 3.3.1, in that

Figure 3-47 Shear failure of a short column in a school building, Tokachi-Oki earthquake, Japan, 1968.

Figure 3-48 Failure of columns in Olive View Hospital, San Fernando earthquake, California, 1971. [*From G. A. Frazier, J. H. Wood, and G. W. Housner,* Earthquake Damage to Buildings; Engineering Features of the San Fernando Earthquake, *EERL 71-02, Earthquake Eng. Res. Lab., California Institute of Technology, Pasadena, Calif., 1971, pp. 140–298.*]

damage is concentrated in the first story because its stiffness is too small compared with that of other stories.

Figure 3-49. An example of column flexural failure at both ends.

Figure 3-50. This frame contained short columns, long columns, and shear walls in the same story. Shear failure of the short columns and shear walls was followed by flexural failure of the long columns (AIJ, 1968).

Figure 3-51. An example of factor 6 in Sec. 3.3.1. The columns failed owing to the torsion caused in the first story by eccentric placement of the shear walls.

Figure 3-52. The extended portion of the school building showed large story drift owing to the pounding effect at the construction joint.

Figure 3-53. An example of factor 5 in Sec. 3.3.1, i.e., shear failure of short members around openings in the wall.

Figure 3-49   Flexural failure of a column: close-up view of Fig. 3-51.

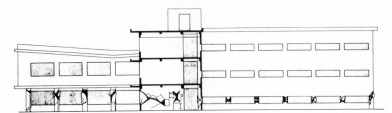

Figure 3-50   Failure of short columns and shear walls followed by the failure of long columns in a technical college, Tokachi-Oki earthquake, Japan, 1968. [*From AIJ*, Report on the Damage Due to 1968 Tokachi-Oki Earthquake, *Tokyo, 1968 (in Japanese).*]

Figure 3-51   Failure of an office building owing to torsional vibration, Miyagiken-Oki earthquake, Japan, 1978.

Figure 3-52   Failure of a high school building owing to pounding at the construction joint, Tokachi-Oki earthquake, Japan, 1968.

Figure 3-53 Failure of walls with openings in an office building, Miyagiken-Oki earthquake, Japan, 1978.

Figure 3-54. An example of simultaneous shear failure of a column and a shear wall (see Fig. 3-31*a*).

Figure 3-55. A precast-concrete factory building has completely collapsed because of nonmonolithic construction and a loss of integrity.

3.4 Behavior of Steel Structures

3.4.1 Introduction

Steel is counted as a good earthquake-resistant material because of its high ductility and large strength-to-weight ratio. Steel structures have performed well under past earthquakes, as described in Sec. 3.4.8. However, a steel structure, which is manufactured from a ductile material, is not always ductile, mainly because of instability and brittle fracture. The following phenomena fall into the category of instability: local buckling of plate elements with large width-thickness ratios, flexural buckling of long columns and braces, lateral-torsional buckling of beams and beam columns, and a P-Δ effect in frames subjected to a large vertical load. Causes of brittle fracture are tension failure at net sections of bolted or riveted connections, fracture of welds subjected to stress concentration, lamellar tearing of plates in which the "through-thickness" strain due to

Figure 3-54 Failure of a shear wall in a school building, Miyagiken-Oki earthquake, Japan, 1978 [*From AIJ*, Report on the Damage Due to 1978 Miyagiken-Oki Earthquake, *Tokyo, 1980 (in Japanese).*]

Figure 3-55 Complete collapse of a precast-concrete factory building, Tangshan earthquake, China, 1976.

weld-metal shrinkage is large and highly restrained, fracture of plates owing to large strains caused by local or flexural buckling, and fatigue caused by cyclic loading with large strain amplitude. If the designer can handle these problems, it is possible to provide a steel structure with sufficient ductility and energy-dissipation capacity.

3.4.2 Local Buckling

A thin-walled steel member which contains a plate element with a large width-thickness ratio is unable to reach its yield strength because of prior local buckling. Even if the yield strength is attained, ductility will be inadequate. A limit thus must be set to the width-thickness ratio. In particular, a more severe limit may well be needed in earthquake-resistant structures with their high ductility requirements than in structures carrying vertical loads only (Popov, Zayas, and Mahin, 1979).

3.4.2.1 Local Buckling under Monotonic Loading   The compressive stress-strain curves of square steel tubes with various width-thickness ratios are schematically illustrated in Fig. 3-56. Curve A drops down owing to local buckling before it reaches the yield stress F_y, and ductility is very small. Curves C and D show sufficient ductility as well as sufficient strength.

The moment-rotation curves in Fig. 3-57 for H-section cantilever beam columns indicate that their strength and ductility are affected by the values of the width-thickness ratio of the flanges (Mitani, Makino, and Matsui, 1977). In the figure M_{pc} is the plastic moment reduced by axial force, and θ_{pc} is the corresponding chord rotation.

3.4.2.2 Local Buckling under Cyclic Loading   Figure 3-58 shows hysteretic load-deflection relationships for cantilever beam columns subjected to a cyclic horizontal load. The specimens are the same as those used in the

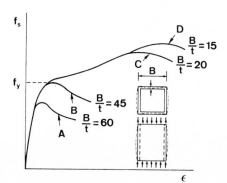

Figure 3-56 Local buckling of square tubes.

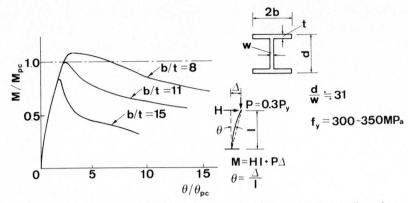

Figure 3-57 Relationships between moment and rotation angle for cantilever beam columns with various values of the width-thickness ratio. [*From I. Mitani, M. Makino, and C. Matsui, Influence of local buckling on cyclic behavior of steel beam-columns,* Proc. Sixth World Conf. Earthquake Eng., *New Delhi,* **3,** *3175–3180 (1977).*]

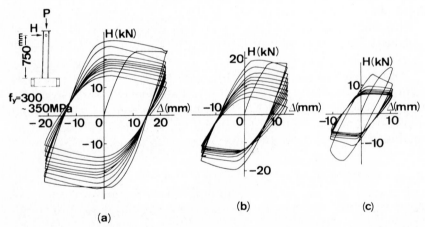

Figure 3-58 Horizontal load-deflection relationships of cantilever beam columns with various values of the width-thickness ratio. *(a) b/t = 8. (b) b/t = 11. (c) b/t = 16.* [*From I. Mitani, M. Makino, and C. Matsui, Influence of local buckling on cyclic behavior of steel beam-columns,* Proc. Sixth World Conf. Earthquake Eng., *New Delhi,* **3,** *3175–3180 (1977).*]

monotonic loading tests shown in Fig. 3-57 (Mitani, Makino, and Matsui, 1977). The specimen in Fig. 3-58a has a width-thickness ratio for the flange equal to 8, which is near the limiting value stipulated in Part 2 of American Institute of Steel Construction (AISC) specifications (see Sec. 4.6.5.1). It is observed that strength and ductility decrease with an increasing width-thickness ratio, and local buckling of the web, following flange buckling, causes further strength degradation.

3.4.3 Beams

3.4.3.1 Behavior under Monotonic Loading   Moment-rotation relationships for H-section beams subjected to uniform bending moment depend on the ratio of laterally unsupported length ℓ_b to the radius of gyration about the weak axis of the section r_y, as schematically shown in Fig. 3-59, where in-plane behavior is illustrated in a and out-of-plane behavior in b. Curve A with a small value of ℓ_b/r_y achieves large rotation with the moment strength maintained at M_p, which is given by

$$M_p = Z_p F_y \tag{3-36}$$

where Z_p = the plastic section modulus and F_y = the yield stress of steel. For curve B, where ℓ_b/r_y becomes a little larger than in curve A, strength is reduced owing to lateral buckling before sufficient rotation is achieved. If ℓ_b/r_y becomes even larger, lateral buckling takes place before moment strength reaches M_p, as shown by curve C.

The relationships between moment and lateral deflection shown in Fig. 3-59b indicate that lateral buckling occurs immediately after moment reaches M_p, but in the case of curve A in-plane and out-of-plane deflections keep increasing without moment reduction because the unsupported length is small.

Test results shown in Fig. 3-60 for beams subjected to uniform moment also indicate similar behavior for beams of various unsupported lengths (Wakabayashi, Nakamura, and Yamamoto, 1970).

Figure 3-61 shows test results for beams subjected to antisymmetrical bending moment. In beams subjected to varying bending moment, the

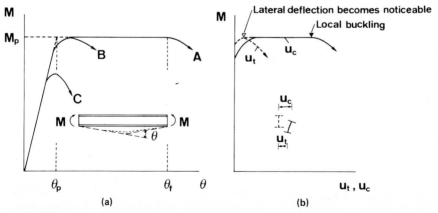

Figure 3-59   Schematic moment-deformation relationships for beams under uniform moment with various values of ℓ_b/r. (*a*) Relationships between moment and rotation angle. (*b*) Relationships between moment and lateral deflection.

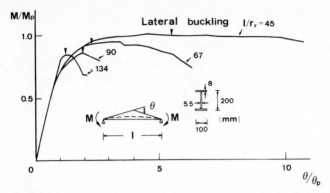

Figure 3-60 Moment-deformation relationships of beams with various values of ℓ/r_y. [*From M. Wakabayashi, T. Nakamura, and H. Yamamoto, Studies on lateral buckling of wide flange beams, rep. 1.* Annuals: Disaster Prevention Res. Inst., *Kyoto University, no. 13A, 365–380 (1970; in Japanese).*]

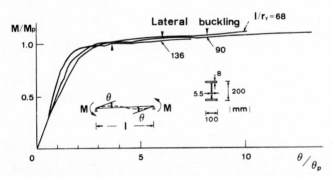

Figure 3-61 Moment-deformation relationships for beams under antisymmetrical moment with various values of ℓ/r_y. [*From M. Wakabayashi, T. Nakamura, and H. Yamamoto, Studies on lateral buckling of wide flange beams, rep. 1,* Annuals: Disaster Prevention Res. Inst., *Kyoto University, no. 13A, 365–380 (1970; in Japanese).*]

yield zone is limited to a short length, and this delays lateral buckling. If ℓ_b/r_y is small, moment strength exceeds M_p because of strain hardening after lateral buckling occurs. The unsupported length ℓ_b which is needed to ensure that the moment is equal to M_p in the large-deformation range is therefore larger for beams subjected to varying bending moment than for beams subjected to uniform moment.

Figure 3-62 shows test results of the relationship between the slenderness ratio ℓ_b/r_y and rotation capacity R of H-shaped beams subjected to uniform or varying bending moments (AIJ, 1975a). The definition of rotation capacity is

$$R = \frac{\theta_f}{\theta_p} - 1 \tag{3-37}$$

where θ_f = the rotation angle at which the curve drops down and θ_p = the theoretical value of the rotation angle when $M = M_p$ (see Fig. 3-59). It is observed that R decreases with increasing ℓ_b/r_y, and the value of R under varying moment is larger than that under uniform moment. Similar results have been shown in several references (Lay and Galambos, 1965; Lay and Galambos, 1967; Galambos and Lay, 1965). To provide a plastic hinge with sufficient rotation capacity for the formation of the plastic collapse mechanism, ASCE-WRC (1971) and AISC (1978) recommend as follows: Members shall be braced at plastic-hinge locations; and the laterally unsupported distance ℓ_b from a braced-hinge location to a similarly braced adjacent point shall not exceed the values given by

$$\ell/r_y = 9480/F_y + 25 \quad \text{MPa} \quad \text{when} \quad 1.0 \geqq M/M_p > -0.5$$

$$\ell/r_y = 9480/F_y \quad \text{MPa} \quad \text{when} \quad -0.5 \geqq M/M_p \geqq -1.0 \tag{3-38}$$

where F_y = the yield stress, MPa; M = the lesser of the moments at the ends of the unbraced segment; and M/M_p = the end moment ratio and is positive when the segment is bent in reverse curvature and negative when bent in single curvature.

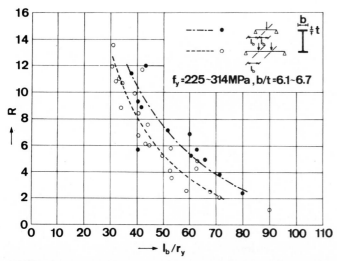

Figure 3-62 Relationship between rotation capacity and the slenderness ratio for beams under uniform and varying bending moments. [*From AIJ*, Guide of the Plastic Design of Steel Structures, *Tokyo, 1975 (in Japanese).*]

The effect of shear on the flexural strength of an H-section beam can be ignored, provided shear intensity is less than $(F_y/\sqrt{3})td_w$, where $t =$ the web thickness and $d_w =$ the web depth.

3.4.3.2 Behavior under Cyclic Loading   In steel members subjected to cyclic loading, strength deterioration is often caused by cracks which develop in a locally buckling plate because of repeated bending (Bertero and Popov, 1965) or by local buckling and/or lateral buckling of the web following local buckling of the flange (Bertero, Popov, and Krawinkler, 1972; Popov, Bertero, and Krawinkler, 1974; Vann, Thompson, et al., 1974). Hysteresis loops for a small rotation amplitude are stable, but strength degradation becomes severe, as shown in Fig. 3-63, when rotation amplitude exceeds a value which is less than half of the rotation capacity under monotonic loading (Takanashi, 1974; Suzuki and Ono, 1977). Plots of rotation capacity versus ℓ_b/r_y for cyclic loading lie below those for monotonic loading. A shorter, laterally unsupported length must therefore be specified for beams subjected to cyclic loading. However, the effects of cyclic loading on rotation capacity have not yet been evaluated quantitatively because of a lack of fundamental information. The restraining effect of a slab attached to a beam is expected to increase rotation capacity, but information on this effect under cyclic loading also is not available.

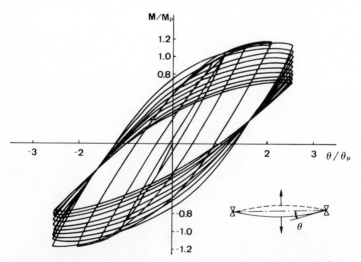

Figure 3-63 Relationships between moment and rotation angle for a beam under cyclic loading. [*From K. Takanashi, Inelastic lateral buckling of steel beams subjected to repeated and reversed loadings,* Proc. Fifth World Conf. Earthquake Eng., *Rome,* **1,** *795−798 (1974).*]

3.4.4 Beam Columns

When a frame is subjected to a horizontal load, the total story shear is either carried entirely by the bracing as shown in Fig. 3-64a or carried by the columns and bracing in combination as shown in Fig. 3-64b. The stress condition for a particular column depends on the load-carrying mecha-

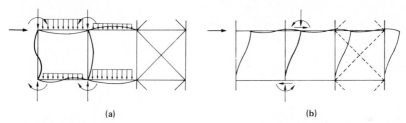

(a)   (b)

Figure 3-64 Beam columns in the frame.

nism. The pattern of plastic-hinge formation around a joint in the frame in Fig. 3-64b is separated into a column-failure type and a beam-failure type, as described in Sec. 4.5.2.3, and the latter is more desirable for steel frames too.

3.4.4.1 Behavior under Monotonic Loading
Moment-curvature relationships for a wide-flange beam column subjected to constant axial force and monotonic bending moment are as shown in Fig. 3-65, where M_p = the full plastic moment under zero axial force and ϕ_y = the yield curvature under zero axial force. The moment rises to M_p and then remains constant with increasing curvature if local buckling does not occur.

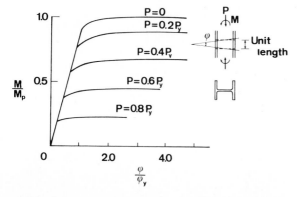

Figure 3-65 Moment-curvature relationships for a wide-flange section.

In addition to failure by flexural yielding, slender beam columns may fail by in-plane instability, caused by compression and flexure, and by local buckling. Figure 3-66 shows moment-rotation relationships for wide-flange beam columns bent about the strong axis, in which lateral-torsional instability is prevented (Driscoll, Beedle, et al., 1965). In the figure ℓ = the length of the beam columns, and r = the radius of gyration about the strong axis of the section. The beam column under antisymmetrical end moment is the most stable, while the ductility of the beam column under symmetrical end moment is the lowest. The strength and ductility of a beam column subjected to moment at one end only lie between these two extremes. It is also observed that instability effects become increasingly important with increasing ℓ/r.

If a wide-flange beam column subjected to compression and bending about its strong axis is not prevented from lateral-torsional instability, ductility is reduced by out-of-plane deflection in addition to in-plane instability as shown in Fig. 3-67 (Wakabayashi, Nakamura, and Naka-shima, 1977). However, out-of-plane instability does not have much effect on the behavior of a wide-flange section of the square type, which is generally used for columns.

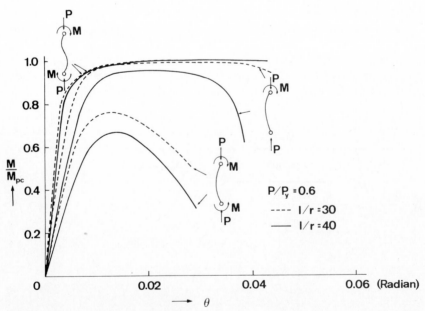

Figure 3-66 Theoretical relationships between moment and rotation angle for beam columns subjected to in-plane instability.

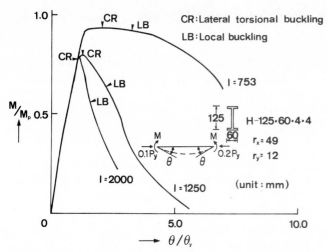

Figure 3-67 Experimental moment–rotation-angle relationships of beam columns subject to lateral-torsional instability. [*From M. Wakabayashi, T. Nakamura, and M. Nakashima, A study on the out of plane inelastic stability of H shaped columns*, Proc. Arch. Inst. Japan Ann. Conv., no. 1, 1399–1400 (1977; in Japanese).]

3.4.4.2 Load-Carrying Capacity   The full plastic moment of a wide-flange section bent about its strong axis is reduced by axial force as follows (ASCE-WRC, 1971):

$$M_{pc} = M_p \qquad \text{for} \qquad 0 < P < 0.15P_y$$
$$M_{pc} = 1.18(1 - P/P_y)M_p \qquad \text{for} \qquad 0.15P_y < P < P_y \tag{3-39}$$

where M_p = the full plastic moment of the section under zero axial force, M_{pc} = the reduced plastic moment, P = the axial thrust, and P_y = the yield axial force. The effect of shear force is usually negligible, as in the case of beams. The strength of a wide-flange beam column which is not prevented from out-of-plane deflection is given by the smaller of the values calculated from the following (AISC, 1978):

$$\frac{P}{P_{cr}} + \frac{C_m M}{(1 - P/P_e)M_m} \leqq 1.0 \tag{3-40}$$

$$\frac{P}{P_y} + \frac{M}{1.18M_p} \leqq 1.0 \qquad \text{and} \qquad M \leqq M_p \tag{3-41}$$

where M = maximum applied moment
  $P_e = \pi^2 EA/(K\ell/r)^2$ = Euler buckling load
  P_{cr} = axial load producing failure in the absence of bending moment, computed for weak-axis buckling

ℓ = actual length in the plane of bending
r = corresponding radius of gyration
K = effective-length factor
C_m = 0.85 for beam columns subjected to joint translation
C_m = $0.6 - 0.4M_1/M_2$ but not less than 0.4 for beam columns braced against joint translation
M_1/M_2 = ratio of the smaller to the larger end moment, which is positive when the member is bent in reverse curvature and negative when it is bent in single curvature

If the P-Δ effect is taken into account in the stress analysis of the whole frame, K may be taken as equal to unity and C_m may be calculated from the formula for beam columns braced against joint translation. M_m designates the lateral buckling moment, which is obtained by treating a given beam column as a beam subjected to uniform bending moment, and its approximate value is calculated as

$$M_m = \left[1.07 - \frac{(\ell_b/r_y)\sqrt{F_y}}{8300} \right] M_p \leqq M_p \tag{3-42}$$

where ℓ_b = the laterally unsupported length, r_y = the radius of gyration about the weak axis, and F_y is in megapascals. For the case of strong axis bending with out-of-plane deflection restrained or weak axis bending,

$$M_m = M_p \tag{3-43}$$

Equation (3-40) indicates that the maximum strength of a beam column is limited by the secondary moment effect which is caused by axial force and excessive deformation, while Eq. (3-41), which is the same as Eq. (3-39), indicates that maximum strength is limited by the formation of a plastic hinge at the end.

3.4.4.3 Behavior under Cyclic Loading
Figure 3-68 shows hysteretic load-deflection relationships for a laterally braced beam column with a compact section subjected to a constant axial force and repeated lateral load (Tanabashi, Yokoo, et al., 1971). Note that the loops enlarge and the maximum strength increases because total compressive strain extends into the strain-hardening range. However, local buckling causes strength degradation in a noncompact section, as shown in Fig. 3-58. The strength of a laterally unbraced beam column initially increases because of the strain-hardening effect, but eventually lateral-torsional buckling causes strength degradation as shown in Fig. 3-69 (Vann, Thompson, et al., 1974). A wide-flange section of the square type, which is strong in torsion, is usually employed as a column section, but strength degradation still takes place under cyclic loading with the deflection amplitude less than

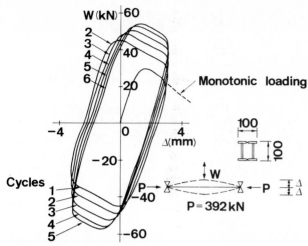

Figure 3-68 Load-deflection relationships for a laterally braced beam column with a compact section subjected to constant axial force and repeated lateral load. [*From R. Tanabashi, Y. Yokoo, et al., Deformation history dependent inelastic stability of columns subjected to combined alternately loading,* RILEM Colloq. Int. Symp., *Buenos Aires (1971).*]

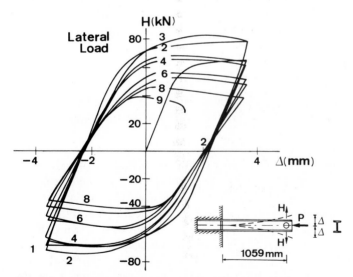

Figure 3-69 Load-deflection relationships for a cantilever beam column subjected to constant axial force and repeated lateral load and failing because of lateral-torsional buckling. [*From W. P. Vann, L. E. Thompson, et al., Cyclic behavior of rolled steel members,* Proc. Fifth World Conf. Earthquake Eng., *Rome, 1, 1187–1193 (1974).*]

the deflection capacity obtained under monotonic loading. Rotation capacity under cyclic loading thus becomes smaller than that under monotonic loading.

3.4.5 Bracing Members

Research on the behavior of bracing under earthquake loading is relatively new (Wakabayashi, 1970; Shibata, Nakamura, et al., 1974; Hanson, 1975; Popov, Takanashi, and Roeder, 1976; Wakabayashi, 1982). Bar bracing is the simplest type and is often used in relatively light one- or two-story building frames. Figure 3-70 shows hysteretic behavior for a bracing bar under repeated loading, where P and δ = the axial force

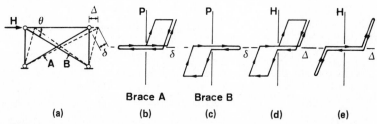

	Brace A	Brace B		
(a)	(b)	(c)	(d)	(e)

Figure 3-70   Hysteresis curves for bar braces. [*From M. Wakabayashi, The behavior of steel frames with diagonal bracings under repeated loading, Proc. U.S.–Japan Semin. Earthquake Eng. with Emphasis on Safety School Build., Sendai, Japan, 328–345 (1970).*]

induced in a brace and corresponding elongation, respectively; and H_B and Δ = the horizontal load carried by braces and horizontal deflection of the frame, respectively. Since each brace can carry only tension as shown in Fig. 3-70b and c, the total horizontal load carried by the bracing is obtained as shown in d by summing the horizontal components of the forces induced in the bars. The hysteresis loop shown in d indicates that the brace dissipates energy only when it experiences newly developed plastic elongation. The brace does not dissipate energy at all if it is subjected to repeated loading under constant deflection amplitude as shown in e, and thus it is said in general that the energy-dissipation capacity of the bracing is less than that of the moment frame.

Hysteresis loops for bracing members which are stockier than the bracing bar become more complex, as shown in Fig. 3-71. Letters indicating various portions of the hysteresis loop in a correspond to the letters in b which show various deflected shapes and loading conditions of the brace. The brace is alternately and repeatedly subjected to rotation at the plastic hinge, which forms because of buckling in compression and plastic

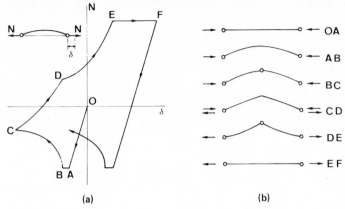

(a)   (b)

Figure 3-71   Hysteresis curves for a brace.

elongation following yielding in tension. The relationship between axial force and axial deformation is theoretically determined from the equilibrium condition, the yield condition, and the associated flow rule, all applied at the plastic hinge (Nonaka, 1973). The boomerang-shaped hysteresis loop becomes thinner and energy-dissipation capacity becomes smaller as the slenderness ratio of the brace becomes larger. Figure 3-72 shows the results of a cyclic-loading test of a single brace under various deformation amplitudes. The shape of the hysteresis loop is similar to that in Fig. 3-71. It is observed that strength degrades owing to load repetition, and the loop stabilizes after a few cycles of loading under the same deformation amplitude (Shibata, Nakamura, et al., 1973). It has been

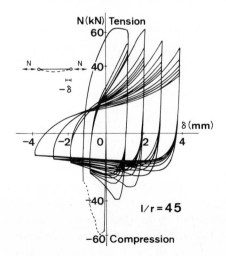

Figure 3-72   Hysteresis loops for a bar. [*From M. Shibata, T. Nakamura, et al., Elastic-plastic behavior of steel braces under repeated axial loading*, Proc. Fifth World Conf. Earthquake Eng., *Rome, 1, 845–848 (1974)*.]

known that bracing members with end conditions other than simple supports can be handled by the effective-length concept (Wakabayashi, Matsui, and Mitani, 1977). It is also reported that the shape of the hysteresis loop varies mainly with the slenderness ratio $K\ell/r$ and that behavior in the large-deformation range may be affected also by the shape of the cross section since local buckling and/or buckling of a single strut of a built-up member may take place (Wakabayashi, Nakamura, and Yoshida, 1977, 1980; Black, Wenger, and Popov, 1980; Jain, Goel, and Hanson, 1980; Maison and Popov, 1980; Astaneh-Asl, Goel, and Hanson, 1982; Gugerli and Goel, 1982). Local buckling often causes cracks in the middle and end portions of the member; the end connection must therefore be detailed so as to avoid stress concentration. In addition, the strength of the end connection must be larger than the yield strength of the bracing member; otherwise, a brittle failure of the connection may occur prior to yielding of the member (Wakabayashi, 1970).

The hysteretic behavior of the brace could be analyzed by the finite-element method (Fujimoto, Wada, et al., 1973), but the plastic-hinge method described in relation to Fig. 3-71 is often employed for the analysis to reduce calculation. This method has been used in the time-history inelastic dynamic analysis of braced frames (Fujiwara, 1980; Igarashi, Inoue, et al., 1972). On the other hand, simpler models of hysteresis loops for the brace have been proposed for the dynamic analysis of braced frames. Such proposals include a multilinear model (Jain and Hanson, 1980) and a model consisting of several mathematically defined curves (Wakabayashi, Nakamura, et al., 1977).

Eccentric bracing systems such as those shown in Fig. 3-73 avoid the reduction in energy-dissipation capacity due to buckling of the brace. They have been investigated and applied in practice in the United States and Japan. The systems illustrated in the figure all ensure large energy-dissipation capacity and good dynamic response (Fujimoto, Aoyagi, et al., 1972; Takeda, 1971; Tanabashi, Kaneta, and Ishida, 1974; Roeder and Popov, 1978; Popov and Bertero, 1980; Popov, 1980a, 1980b).

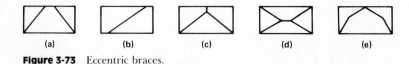

   (a)   (b)   (c)   (d)   (e)

Figure 3-73   Eccentric braces.

3.4.6 Connections

3.4.6.1 Load-Carrying Capacity   Failure of a welded rigid beam-to-column connection may occur by yielding or fracture as a result of high local stress as shown in Fig. 3-74a or, alternatively, by shear yielding of the

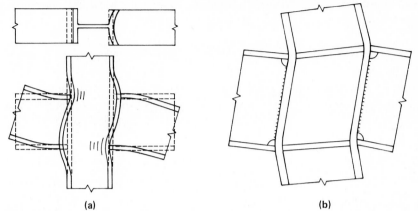

Figure 3-74   Failure of beam-to-column connections. *(a)* Failure due to local stress. *(b)* Shear yield of the connection panel.

connection panel as shown in *b*. Local stress, developed by the compression and tension forces delivered from the beam flanges, can bring about two types of failure: (1) crippling of the column web due to the compression force delivered from the beam compression flange; and (2) excessive flexural deformation of the column flange followed by fracture of the flange weld in the vicinity of the column web, caused by the tension force delivered from the beam tension flange.

In the first case the compressive flange force at yield, A_fF_y, is assumed to spread at an angle of 1:2.5 and to distribute itself uniformly into the column web, as shown in Fig. 3-75. If the column web is not to yield when the beam flange yields, then we have (AISC, 1969)

$$F_yt_w(t_b + 5k_c) \geqq A_fF_y \tag{3-44}$$

where k_c = the distance from the outer face of the flange to the web toe of the column, A_f = the area of the beam flange, t_b = the thickness of the

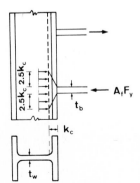

Figure 3-75   Local stress in the column web.

beam flange, and t_w = the thickness of the column web. To avoid web buckling, the width-thickness ratio of the column web must satisfy

$$\frac{d_c}{t_w} \leqq \frac{473}{\sqrt{F_y}} \qquad \text{MPa} \tag{3-45}$$

where d_c = the column-web depth between fillets and F_y is in megapascals.

For the second type of failure, the thickness of the column flange which provides sufficient flexural stiffness to resist the force from the tension flange of the beam is as follows:

$$t_c \leqq 0.4\sqrt{A_f} \tag{3-46}$$

where t_c = the thickness of the column flange. Stiffeners are not needed on the column web at the beam compression flange if Eqs. (3-44) and (3-45) are both satisfied; they are not needed opposite the beam tension flange if Eq. (3-46) is satisfied. A new AISC specification (1978) prescribes equations similar to Eqs. (3-44) through (3-46). In the discussions above, it is assumed that the yield stresses of the beam and the column are both equal to F_y. If the web-crippling stress and the shear stress in the connection panel are both large, the formula for combined stress must be used to design the connection, since the effect of the combined stress is not taken into account in Eqs. (3-44) and (3-45) (ATC-3, 1978).

The shear force V_p acting on the connection panel can be calculated in the same manner as in the reinforced-concrete connection (Fig. 3-37), that is, by referring to Fig. 3-76, from Eq. (3-30). The yield shear strength of the panel is given by

$$V_{py} = \frac{F_y}{\sqrt{3}} A_w \tag{3-47}$$

where $A_w = t_w j_c$ and t_w = the panel thickness. The depths of the panel j_b and j_c are taken equal to $0.95d_b$ and $0.95d_c$, respectively, where d_b and d_c are

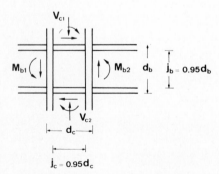

Figure 3-76 Shear and moment on a connection.

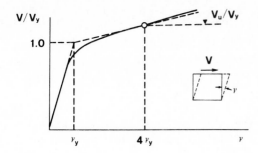

Figure 3-77 Relationship between shear force and shear deformation for a connection panel under monotonic loading.

the depths of the beam and the column, respectively. Equation (3-47) is based on the assumption that the shear stress is uniformly distributed over the length j_c, even though the shear stress is at a maximum at the center of the panel, and also on the assumption that the column axial force and the bending moment are both carried by the column flanges only. A_w designates that area of the web panel of a wide-flange column, and half a cross-sectional area is taken for A_w in the case of a circular tube column or a square box column with constant wall thickness. The forces acting on the connection at the onset of panel yielding can be determined from Eqs. (3-30) and (3-47). Equation (3-30) can also be applied to the connection of the exterior column with one beam only, by equating one M_b term to 0.

Figure 3-77 shows the relation between shear force and shear distortion for a panel under monotonic loading. The slope of the curve changes in the vicinity of the yield strength V_y as given by Eq. (3-47) and becomes 3 to 8 percent of the elastic slope. Strength increases well beyond yield strength, and the ductility factor reaches 30 to 40 (Kato, 1982).

Extra strength beyond yield strength is provided by the following effects:

1. Resistance of the boundary elements, i.e., column flanges and diaphragms

2. Resistance of the beam and column webs adjacent to the connection panel

3. Strain hardening of the panel (Kato, 1982; Krawinkler and Popov, 1982; Popov, Bertero, and Chandramouli, 1975)

Since the extra strength is very large and there is no evident reason why the panel must not yield prior to yielding of the surrounding members, it has been proposed that design of the panel be based on a higher strength than yield strength. For example, Kato (1982) and the AIJ (1979) recommend that the shear term [the second term on the right-hand side of Eq. (3-30] be neglected and that F_y in Eq. (3-47) be multiplied by ⅓. In

Krawinkler and Popov (1982), the strength at $4\gamma_y$ as shown in Fig. 3-77 is recommended as the basis of the design of the panel.

When panel thickness is not sufficient, the panel is strengthened by doubler plates, the most effective way being to weld the plates directly to the column web. Effectiveness is reduced if the doubler plates are welded to the column flanges at some distance from the column web (Krawinkler and Popov, 1982).

The strength of the steel material for the "through-thickness" direction is nearly equal to its strength for the direction parallel to rolling, but through-thickness ductility is small (Tall Building Committee 43, 1979). When welds are made in highly restrained beam-to-column connection as shown in Fig. 3-78a, the localized through-thickness strain in the column flange due to weld-metal shrinkage exhausts ductility, which results in a brittle fracture in the form of lamellar tearing. In addition, cracks in the column flange begin from the root of the beam flange welds or the end of the scallop as shown in Fig. 3-78b. To avoid these cracks and lamellar tearing, weld detail should be such that weld shrinkage occurs in the rolling direction and restraint is minimized.

3.4.6.2 Behavior under Cyclic Loading   Cyclic bending tests of cantilevers connected to rigid stub columns (Popov and Pinkney, 1968, 1969a, 1969b; Popov and Stephan, 1972) reveal the following characteristics of connections:

1. Hysteresis loops of fully welded connections are stable and spindle-shaped as shown in Fig. 3-79.

2. Welded flange and bolted web connections as shown in Fig. 3-80a are not fully rigid owing to bolt slippage, but their hysteretic behavior in general is similar to that of the fully welded connection.

3. For connections with welded flange splices, as shown in Fig. 3-80b, cracks form early in the splice plate at the end of the fillet weld, and hence ductility is smaller than in cases 1 and 2.

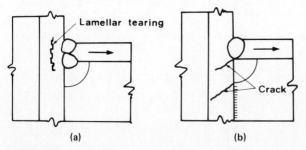

Figure 3-78   Local failure. *(a)* Lamellar tearing. *(b)* Cracks.

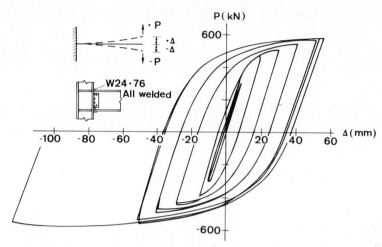

Figure 3-79 Load-deflection curves for a connection. [*From H. Krawinkler and E. P. Popov, Seismic behavior of moment connections and joints,* J. Struct, Div., Am. Soc. Civ. Eng., *108(ST-2), 373–391 (1982).*]

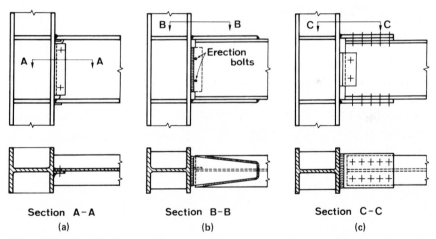

Section A–A   Section B–B   Section C–C

(a)   (b)   (c)

Figure 3-80 Details of connections. [*From H. Krawinkler and E. P. Popov, Seismic behavior of moment connections and joints,* J. Struct. Div., Am. Soc. Civ. Eng., *108(ST-2), 373–391 (1982).*]

4. Connections with bolted flange and web splices as in Fig. 3-80c show hysteresis loops of the slip type as indicated in Fig. 2-37 because of bolt slippage (Krawinkler and Popov, 1982).

Many tests of connection-panel zones under cyclic loading have been carried out (Naka, Kato, et al., 1969; Krawinkler, Bertero, and Popov, 1971; Bertero, Popov, and Krawinkler, 1972; Krawinkler, Bertero, and

Popov, 1975). It is concluded from the test results that a carefully detailed connection shows stable, spindle-shaped hysteresis loops and that ultimate strength is much higher than yield strength. Strength degradation is scarcely observed in such connections even if diagonal buckling occurs in the panel owing to the shear force.

3.4.6.3 Influence of Shear Distortion of Connection Panels on Frame Behavior
The elastic deformation of a frame may be underestimated by analysis if shear distortion of the connection panel is not taken into account (Bertero, Krawinkler, and Popov, 1973; Bertero, 1969). However, the error is not in fact very large because it is often assumed in the analysis that the size of the connection is negligible and the frame is represented by member centerlines; this results in greater member lengths and thus in reduced frame stiffness. This assumption compensates for the error caused by neglecting shear distortion in the panel (Kato, 1982; Kato and Nakao, 1973).

3.4.7 Systems

3.4.7.1 Unbraced Frames   Horizontal load-deflection relationships for multistory frames subjected to constant vertical load and monotonic horizontal load are shown in Fig. 3-81 (Wakabayashi, Nonaka, and Morino, 1969). The strength of a frame under zero vertical load continues to rise beyond the simple plastic collapse load H_p because of the strain-hardening effect. Maximum strength of the frame decreases with increasing vertical load as a result of the P-Δ effect, i.e., the effect of the overturning moment produced by the vertical load P and the horizontal deflection Δ. Frame

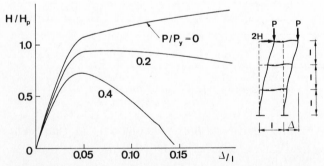

Figure 3-81 Horizontal load-deflection relationships for frames under constant vertical load and monotonic horizontal load. [*From M. Wakabayashi, T. Nonoka, and S. Morino, An experimental study on the inelastic behavior of steel frames, with a rectangular cross-section subjected to vertical and horizontal loading, Bull. Disaster Prevention Res. Inst., Kyoto University, 18(145), 65–82 (1969)*].

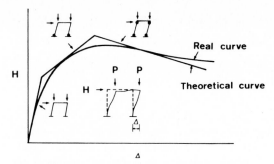

Figure 3-82 Horizontal load-deflection relationship for a portal frame under constant vertical load and monotonic horizontal load.

behavior is discussed in Arnold and Adams (1968), Wakabayashi, Nonaka, and Matsui (1969), Wakabayashi, Nonaka, and Morino (1969), and Wakabayashi, Matsui, et al. (1974a, 1974b).

A relatively simple method of analyzing the load-deflection relation of a frame is called *unit shape factor analysis,* which assumes that the inelastic deformation is concentrated at the plastic hinge and that all other regions remain elastic. The secondary effect of axial force is considered in member stiffness, and incremental relations between horizontal load and deflection are repeatedly determined for the frame in which the plastic hinges are replaced by real hinges, as shown in Fig. 3-82. As the strain-hardening effect is neglected in this method of analysis, the calculated load-deflection curve lies below the real curve in the large-deflection range (Vogel, 1963; Yarimci, Parikh, et al., 1965; Wakabayashi, 1965). The subassemblage method has also been developed for manual computation of the load-deflection relation of the frame. A particular level is separated from the frame at the inflection points in the columns above and below the floor, and a further subdivision is made into subassemblages consisting of a column and adjacent beams. The horizontal force-sway curve for each subassemblage is determined by using load-deformation curves. The load-deflection curve for the whole frame can be obtained by adding the individual subassemblage curves (Driscoll, Beedle, et al., 1965; Daniels and Lu, 1966; de Buen, 1969).

It is known that frame instability is influenced by the slenderness ratio of the columns, the ratio of the working axial force to the yield axial force of the columns, and the beam-stiffness ratio. An attempt has been made to express the ductility factor μ as a function of these parameters (Sakamoto and Miyamura, 1966). In Fig. 3-83, experimentally obtained ductility factors for portal frames are shown by circles, with the curve expressed by

$$\mu = \frac{\delta_{cr}}{\delta_y} = 0.7 + \frac{0.3\pi \ \sqrt{E/F_y}}{\sqrt{P/P_y}K\ell/r} \tag{3-48}$$

where P_y = yield axial force of the column
$K\ell$ = effective column length
r = radius of gyration of the column section
E = Young's modulus
F_y = yield stress

Other symbols are defined in the inset in Fig. 3-83. Ductility decreases with increasing $K\ell/r$ and P/P_y.

Figure 3-84 shows horizontal load-deflection relationships for full-scale portal frames subjected to a cyclic horizontal load (Wakabayashi, Matsui, et al., 1974a, b). Hysteresis loops in *a* are spindle-shaped, since no vertical load is applied. The negative slope which appears in the loops in *b* after maximum strength has been attained is caused by the P-Δ effect. Similar behavior is observed in the tests of full-scale three-story frames reported in Carpenter and Lu (1969).

Figure 3-85 schematically illustrates the hysteretic relationship between an alternately repeated horizontal load H and displacement δ for a portal frame. Plastic deformation takes place in the loading process when the load exceeds the maximum load attained in previous cycles, i.e., in loading path $B'C$. The amount of plastic deformation accumulated in path $B'C$ is equal to that in path BC'' in the monotonic loading curve. The total amount of plastic deformation accumulated in the hysteresis loops

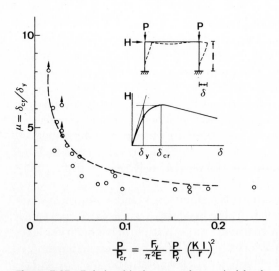

Figure 3-83 Relationship between the vertical load and ductility for a portal frame. [*From J. Sakamoto and A. Miyamura, Critical strength of elastoplastic steel frames under vertical and horizontal loadings,* Trans. Arch. Inst. Japan, *no. 124, 1–7 (1966; in Japanese).*]

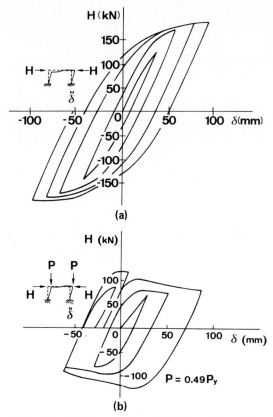

(a)

(b)

Figure 3-84 Horizontal load-deflection relationships for portal fames. *(a)* Without vertical load. *(b)* With vertical load. [*From M. Wakabayashi, C. Matsui, et al., Inelastic behavior of full scale steel frames with and without bracings,* Bull. Disaster Prevention Res. Inst., *Kyoto University, 24, part 1, no. 216, 1−23 (1974); Inelastic behavior of steel frames subjected to constant vertical and alternating horizontal loads,* Proc. Fifth World Conf. Earthquake Eng., *Rome, 1, 1194−1197 (1974).*]

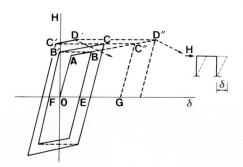

Figure 3-85 Theoretical hysteretic model of the load-deflection relationship for steel members and frames. [*From B. Kato and H. Akiyama, Theoretical prediction of the load-deflection relationship of steel members and frames,* Symposium on Resistance and Ultimate Deformability of Structures Acted on by Well-Defined Repeated Loads: Preliminary Report, *International Association of Bridge and Structural Engineering, Lisbon Symposium,* 23−28 (1973).]

$OAB\text{-}B'C\text{-}C'D$ is thus equal to the amount of plastic deformation occurring in the monotonic loading curve $OABC''D''$. Let us suppose that local buckling occurs at D'' in a monotonically loaded frame; then local buckling occurs at D in the same frame subjected to repeated loading (Kato and Akiyama, 1973).

3.4.7.2 Braced Frames   Hysteresis loops obtained from tests of full-scale braced frames are shown in Fig. 3-86a, and a typical loop is compared with the theoretical loop in b (Wakabayashi, Matsui, et al., 1974a,b). Dynamic analysis reveals that, for multistory braced frames in which bracing resists more than 50 percent of the total horizontal load, plastic deformation is concentrated in a particular story when the intensity of input ground motion exceeds a certain level. This is true because the energy-dissipation capacity of braced frames is generally small compared with that of unbraced frames (Sakamoto and Kohama, 1973).

3.4.8 Earthquake Damage

The San Francisco (1906) and Kanto (1923) earthquakes demonstrated the good earthquake-resistant capacity of steel frames. Excellent performance of steel high-rise buildings, including the Latin-American Tower, was observed during the Mexico City earthquake (1957). Since then problems have not occurred in steel frames in the Anchorage (1964), Caracas (1967), Tokachi-Oki (1968), and San Fernando (1971) earthquakes, provided that design, fabrication, and erection were adequate. However, damage to finishing materials and nonstructural members has been often reported (de Buen, 1980). In the Miyagiken-Oki earthquake (1978), medium-scale building frames were damaged because of the fracture of bracing connections. Figure 3-87 shows that fracture of braces caused the complete failure of the first story of a two-story braced-frame warehouse.

3.5 Behavior of Composite Structures

3.5.1 Introduction

A composite structure possesses the properties of both concrete and steel, and by appropriate design it is possible to provide good earthquake resistance in such structures. High-rise buildings of composite construction showed good earthquake-resistant capacity under the Kanto earthquake (1923) as compared with ordinary reinforced-concrete structures.

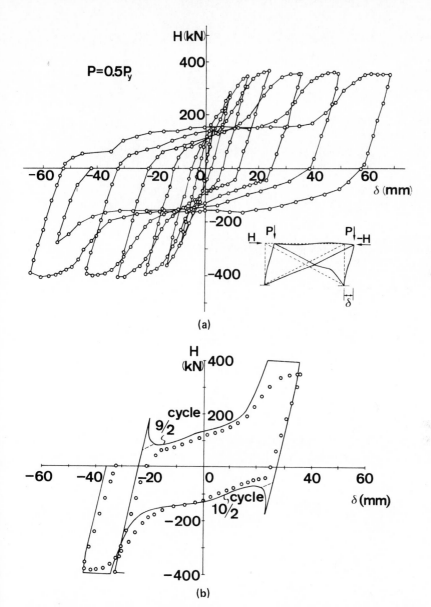

Figure 3-86 Load-deflection relationships for a braced full-scale portal frame under a constant vertical load and an alternately repeated horizontal load. *(a)* Experimental curves. *(b)* Comparison of experimental and theoretical curves. [*From M. Wakabayashi, C. Matsui, et al., Inelastic behavior of full scale steel frames with and without bracings,* Bull. Disaster Prevention Res. Inst., *Kyoto University,* **24**, *part 1, no. 216, 1–23 (1974); Inelastic behavior of steel frames subjected to constant vertical and alternating horizontal loads,* Proc. Fifth World Conf. Earthquake Eng., *Rome,* **1**, *1194–1197 (1974).*]

Figure 3-87   Collapse of a warehouse, Miyagiken-Oki earthquake, Japan, 1978. *(above)* Outside view. *(below)* Failure of a joint in the brace in the second story.

Since then the encased structural system, a form of composite construction, has been employed in Japan for most building frames higher than seven stories.

The flexural behavior of an encased member is similar to that of a reinforced-concrete member until maximum strength is attained; however, it possesses larger ductility, since the steel can resist forces after the concrete crushes, provided that the width-thickness ratios of the steel plates are sufficiently small. As to shear resistance, an encased member with full-web steel shows ductile behavior even when it fails in shear.

Concrete-filled steel-tube columns with small diameter-thickness ratios show ductile behavior under flexure since the steel tubes confine the concrete. Furthermore, strength and ductility are scarcely affected by even relatively large shear forces.

An unencased composite beam is not a very good earthquake-resistant component, since only reinforcement in the concrete slab resists the negative bending moment, even though the slab works effectively against the positive bending moment. The connection between the steel and the concrete slab is important if the diaphragm action of the slab is to be relied on.

Thorough surveys on composite structures are given by Viest (1974), Wakabayashi (1974), Tomii, Matsui, and Sakino (1974), and Tall Building Committee A41 (1979). In a so-called mixed structure which consists of several different types of structural systems, failure often occurs in the connections between two members of different structural type: a steel beam and a reinforced-concrete column, a reinforced slab and a masonry wall.

3.5.2 Concrete-Encased-Steel Members

3.5.2.1 Flexural Behavior   Moment-curvature relationships for concrete-encased-steel members subjected to constant axial force and monotonic bending moment are shown in Fig. 3-88, which indicates that ductility is reduced with increasing axial force, as in the case of reinforced-concrete members (Wakabayashi, 1974).

Hysteresis loops for encased-steel members subjected to constant axial force and cyclic bending moment with shear force are shown in Fig. 3-89. The shape of the loops for a member with zero axial force is more like a spindle than in the case of a reinforced-concrete member as shown in Fig. 3-14b, and ductility decreases with increasing axial force, which is similar to the behavior of a reinforced-concrete member (Wakabayashi and Minami, 1976).

Ultimate flexural strength of an encased member can be obtained by

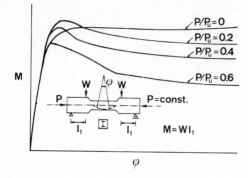

Figure 3-88 Moment-curvature relationships for encased members under constant axial force and monotonic bending moment. [*From M. Wakabayashi, Steel reinforced concrete, elastic plastic behavior of members, connections and frames,* Proc. Nat. Conf. Planning Des. Tall Build., *Tokyo, 3, 23–36 (1974).*]

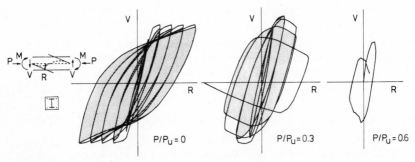

Figure 3-89 Relationships between moment and rotation angle for encased members under constant axial force and alternately repeated bending moment with shear. [*From M. Wakabayashi and K. Minami, Experimental studies on hysteretic characteristics of steel reinforced concrete columns and frames,* Proc. Int. Symp. Earthquake Struct. Eng., *University of Missouri–Rolla, St. Louis, 1, 467–480 (1976).*]

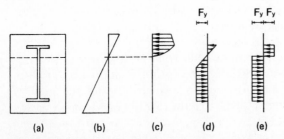

Figure 3-90 Concrete-encased member at the ultimate state. *(a)* Cross section. *(b)* Strain. *(c)* Stress in concrete. *(d)* Actual stress in steel. *(e)* Assumed stress in steel.

equilibrium of internal forces, which, as shown in Fig. 3-90, are calculated from prescribed stress-strain relationships for concrete and steel and from the assumption that plane sections remain plane after deformation occurs. This approach is the same as that employed in the analysis of reinforced-concrete members, and theoretical strength agrees well with experimentally determined strength (Wakabayashi, 1974). To simplify the calculation, the rather complicated stress distribution in the steel shown in Fig. 3-90d is often replaced by the fully plastic distribution in e. The error in ultimate strength caused by this simplification is small (Wakabayashi, 1974).

Another method for strength calculation is the so-called method of superposed strengths, in which the ultimate strength of the section of an encased member is taken to be the sum of the ultimate strength of the reinforced-concrete section plus the strength of the fully plastic steel section. This method involves very simple calculations, and the error involved is still small compared with that of the method based on the assumption that plane sections remain plane (Nakamura and Wakabayashi, 1976). According to the plastic theory, the superposition method gives a lower-bound value for a member composed of very ductile materials. However, this method yields errors on the unsafe side for composite columns because the concrete is not sufficiently ductile. Such an error can be compensated for by reducing the compressive strength of concrete. Thus, in Japan this method has been recommended in the codes and is widely used (AIJ, 1975b). The superposed strength is written in the form

$$P = {}_sP + {}_rP$$
$$M = {}_sM + {}_rM \tag{3-49}$$

where P and M designate the axial force and the bending moment, respectively, and subscripts s and r indicate forces carried by the steel and the reinforced-concrete sections, respectively. The ratios of forces to be carried by the two component sections, i.e., ${}_sP/{}_rP$ and ${}_sM/{}_rM$, are rather arbitrarily chosen. In other words, the superposition indicated by Eq. (3-49) can be made so that the combined strength of P and M becomes the maximum.

Figure 3-91 illustrates the superposition of individual strength-interaction curves for a steel section (marked S) and a reinforced-concrete section (RC), both being shown as dashed lines. The superposed strength given by Eq. (3-49) is represented by the solid line ACF, which is an envelope curve obtained by moving one of the S and RC curves so that its origin stays on the other curve. The dash-and-dot line $ACDF$ is a multilinear approximation of the curve ACF, with the coordinates of points A, C, D, and F defined by the following quantities:

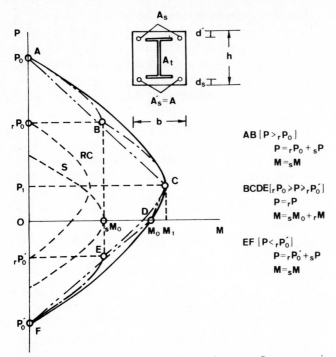

Figure 3-91 Moment−axial-force interaction curve for a composite section by the method of superposition.

$$P_0 = A_t F_y + A_{st} f_y + bh f'_c$$
$$P'_0 = A_t F_y - A_{st} f_y$$
$$M_0 = Z_p F_y + A_s f_y (h - d' - d_s) \qquad\qquad (3\text{-}50)$$
$$P_1 = bh f'_c / 2$$
$$M_1 = Z_p F_y + A_s f_y (h - d' - d_s) + bh^2 f'_c / 8$$

where A_{st} = total area of the main reinforcement
                  A_s = area of the main reinforcement on one side
                  A_t = area of the encased steel
                  Z_p = plastic section modulus of the encased steel
f_y and F_y = yield stresses of the main reinforcement and the encased steel, respectively

Other symbols appearing in Eq. (3-50) are defined in the inset in Fig. 3-91. The rectangular concrete stress distribution shown in Fig. 3-17d with $\beta_1 = 1$ is assumed for deriving Eq. (3-50). f'_c designates the compressive strength of concrete, which is calculated by multiplying the specified compressive strength of concrete by the factor k_3, given as

$$k_3 = 0.85 - 2.5 {}_s p_c \qquad\qquad (3\text{-}51)$$

where $_s p_c$ = the ratio of the compression flange area to the total area of the concrete. This reduction in the compressive strength of concrete compensates for the unsafe error involved in the superposition of strengths as already discussed, and it is applicable only to the concrete in the column. Equation (3-49) gives the generalized strength by superposition, and Eq. (3-50) provides an approximation. Another approximation, shown as the dash-and-dot curve *ABCDEF*, is often used. Mathematical expressions for this curve are given in Fig. 3-91, where

$$_r P_0 = A_{st} f_y + bh f'_c$$
$$_r P'_0 = - A_{st} f_y \tag{3-52}$$

It is clear from the figure that when $_r P_0 \geqslant P \geqslant {}_r P'_0$, the steel section alone carries the moment $_s M_0$ ($_s P = 0$), and when $P > {}_r P_0$ or $_r P'_0 > P$, the reinforced-concrete section alone carries the axial force $_r P_0$ or $_r P'_0$ ($_r M = 0$). The superposed strength of a concrete-encased-steel section is easily calculated if the individual ultimate strengths of the components are determined beforehand.

Since the ductility of an encased column decreases with increasing axial force as shown in Figs. 3-88 and 3-89, the AIJ Standard (AIJ, 1975*b*) recommends that the axial force on an encased column, for which large ductility is required under earthquake loading, shall not exceed P_{cr} given by

$$P_{cr} = \frac{1}{3} (bh f'_c + 2A_t f_y) \tag{3-53}$$

At the ultimate state, the concrete is crushed and the longitudinal reinforcement buckles, but the steel may still carry the load without buckling. Thus, the contribution of the steel is counted in Eq. (3-53) on the premise that the width-thickness ratios of the steel are kept below the limiting values specified in the allowable stress design standards for steel structures.

The ductility of an encased member is related to the interaction between concrete reinforcement and steel. Concrete surrounded by steel flanges can sustain large strain because of the confining action of the flanges. On the other hand, concrete covering steel on the compression side fails in a brittle manner owing to the combined effect of its own buckling and the lateral force induced by the buckling of the steel. Thus, the compressive-concrete cover spalls off at a limited compressive strain, and this causes strength to deteriorate. Spalling of concrete may be prevented by hoop ties, but quantitative relations between the quantity of transverse reinforcement and its confining effect have not yet been evaluated for this particular situation.

3.5.2.2 Shear Behavior
The behavior of an encased column failing in shear is much better than that of a reinforced-concrete column even

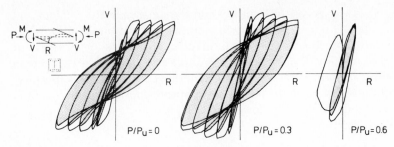

Figure 3-92 Shear-force–rotation-angle relationships of encased beam columns failing in shear under constant axial force and alternately repeated bending moment with shear. [*From M. Wakabayashi and K. Minami, Experimental studies on hysteretic characteristics of steel reinforced concrete columns and frames,* Proc. Int. Symp. Earthquake Struct. Eng., *University of Missouri–Rolla, St. Louis, 1, 467–480 (1976).*]

though the flexural behavior of both columns is rather similar. Figure 3-92 shows hysteresis loops for an encased column failing in shear (Wakabayashi and Minami, 1976); the spindle-shaped loops indicate better ductility and less strength degradation than in the case of a reinforced-concrete column as shown in Fig. 3-23. When the encased member is subjected to cyclic loading under large deformation amplitude, the bond between the steel and the concrete is broken, and the steel member and the reinforced-concrete member then resist the shear force individually. This must be taken into account in practical design.

The shear strength V_0 of an encased column is calculated by superposition as follows:

$$V_0 = {}_{rc}V_0 + {}_sV_0 \tag{3-54}$$

where ${}_sV_0$ = the yield shear strength of the steel member and ${}_{rc}V_0$ = the ultimate shear strength of the reinforced-concrete member. The value of ${}_sV_0$ is taken as the smaller of the yield shear strength of the steel web and the shear force calculated by assuming that the yield moment is attained at both ends of the steel member. In calculating ${}_{rc}V_0$, the splitting failure of the layer parallel to the longitudinal axis of the member as shown in Fig. 3-93 must be taken into consideration in addition to ordinary shear failure. Splitting failure is caused by shear failure of the concrete along the plane shown by the dashed line in the figure, which in turn is the result of a bond breakdown between the outer surface of the steel flange and the concrete. Shear strength at the onset of the splitting failure is given by

$$_{rc}V_0 = r^d(\alpha b' f'_c + \beta f_s A_v/s) \tag{3-55}$$

where parameters α and β are determined as $\alpha = 0.15$ and $\beta = 1$, based on experimental observations (Minami and Wakabayashi, 1974; Wakabayashi, 1976). The value of $_{rc}V_0$ is thus taken as the smaller of the values calculated by Eqs. (3-55) and (3-23).

3.5.3 Concrete-Filled-Steel Tubes

The strength of a concrete-filled-steel tube subjected to compression and bending can be calculated in the same manner as for an encased column, on the assumption that plane sections remain plane after deformation occurs. The stress blocks for the steel and the concrete may be assumed to be rectangular to simplify the calculation. The method of superposed strengths is also applicable and has recently been extended to the calculation of strengths of long beam columns (Wakabayashi, 1977a). Concrete-filled-steel tubes subjected to compression and bending are generally more ductile than encased beam columns. A concrete-filled tube has better resistance to local buckling than an empty tube, but the diameter-thickness ratio should still be kept below a limiting value.

In the case of concrete-filled-steel tubes subjected to compression, bending, and shear, external forces are resisted by the steel tube and the diagonal concrete compression strut (arch mechanism; see Fig. 3-27). True shear failure therefore does not occur even in relatively short members, and flexural strength can be calculated without taking shear into account. The hysteresis loops shown in Fig. 3-94 for short beam columns with concrete-filled-steel tube sections indicate large ductility, large energy-dissipation capacity, and little strength degradation (Wakabayashi and Minami, 1980b).

3.5.4 Unencased Composite Beams

The use of unencased composite beams is effective in increasing beam stiffness and decreasing deflection of structural frames. However, a com-

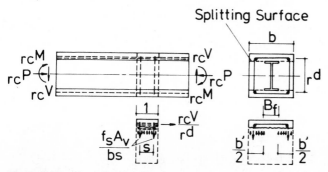

Figure 3-93 Shear-bond failure mechanism.

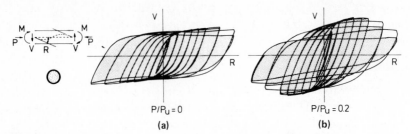

Figure 3-94 Relationships between shear force and rotation angle for concrete-filled tubes under constant axial force and alternately repeated bending moment with shear. *(a)* $P/P_u = 0$. *(b)* $P/P_u = 0.2$. [*From M. Wakabayashi and K. Minami, Experimental study on the hysteretic characteristics of composite beam-columns, in B. Kato and L.-W. Lu (eds.),* Developments in Composite and Mixed Construction, *Proc. U.S.A.–Japan Seminar on Composite Structures and Mixed Structural Systems, Gihodo Shuppan Co., Tokyo, 1980, pp. 197–211.*]

posite beam resists positive moment in cooperation with a compressive-concrete slab, and only the steel reinforcement in the slab works against the negative moment. The composite beam is therefore not very effective for earthquake loading where positive and negative moments occur repeatedly (Kato and Uchida, 1974). In addition, steel reinforcement in the slab adjacent to the exterior column is not securely anchored to the column, and thus the flexural strength of such a composite beam is merely the same as that of a bare-steel beam (Daniels and Fisher, 1970).

3.5.5 Composite Shear Walls

In composite construction, concrete-encased-steel shear walls are often used in addition to the reinforced-concrete walls discussed in Sec. 3.3.5 to ensure high ductility. The strength of a reinforced-concrete shear wall starts to decrease at a story drift angle of about 0.004 as shown in Fig. 3-31. On the other hand, Fig. 3-95 indicates that the deformation capacity of a concrete-encased-steel truss is increased to about 0.01 (Tall Building Committee A41, 1979). In the case of a concrete-encased ribbed-steel sheet, local buckling occurs at a drift angle of about 0.004, but this does not cause much strength reduction: deformation capacity is as much as 0.02 to 0.03 (Makino, Ozaki, and Hirosawa, 1965). A study of a concrete-encased truss with H-steel members has also been reported (Tall Building Committee A41, 1979).

3.5.6 Connections

3.5.6.1 Connection of Encased-Steel or Bare-Steel Beam to Encased Column   In connecting a concrete-encased-steel beam to a column, local compression stress in the column web, induced by the beam compression

flange, is reduced by the surrounding concrete, as compared with the bare-steel connection. However, the local tension stress induced by the beam tension flange is not so much reduced that a stiffener is not needed to protect the web.

In contrast with local stresses, the shear force carried by the steel panel is substantially reduced because of the surrounding concrete. The external shear force V_p acting on the panel is calculated in the same manner as for a reinforced-concrete connection [Eq. (3-30)]. The hypothesis that the strength of an encased connection panel is the sum of the strength of the steel panel plus the strength of the reinforced-concrete panel shows good agreement with test results (Wakabayashi, 1974, 1976). Thus, in view of Eqs. (3-31) and (3-47), the strength is given by

$$V_p = v_c t_{pe} j_c + 0.5 f_y A_{vj} j_c / s + A_w F_y / \sqrt{3} \tag{3-56}$$

By equating Eq. (3-30) to Eq. (3-56), the required value of A_w can be obtained, and hence the panel thickness. In Eq. (3-56), v_c is conservatively given by $v_c = 0.3 f_c'$ instead of Eq. (3-32). The effective thickness of the concrete panel t_{pe} is given by the average of beam and column widths. In the case of the bare-steel beam connected to the concrete-encased-steel column, t_{pe} is taken to be equal to half the width of the column.

Figure 3-96 shows hysteresis loops for a concrete-encased-steel cruciform frame in which the connection panel fails in shear. The load corresponding to shear failure of the panel, H_{ps}, is calculated from Eqs. (3-30) and (3-56) (Wakabayashi, Nakamura, and Morino, 1973). It is observed that both ductility and energy-dissipation capacity are large, and little strength degradation occurs.

3.5.6.2 Connection of Steel Beam to Concrete-Filled Column   In a connection between a bare-steel beam and a concrete-filled-steel tube column, a

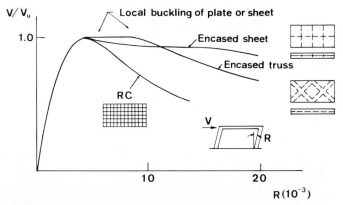

Figure 3-95 Idealized envelope curves of the load-deformation relationships of shear walls failing in shear under repeated loading.

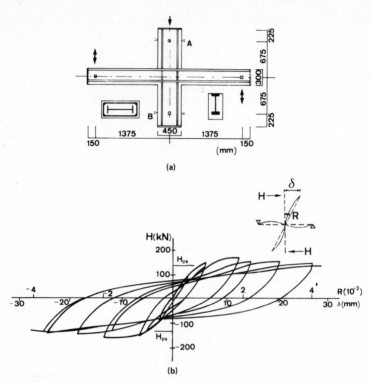

Figure 3-96 Load-deformation relationships for a cruciform frame with the connection panel failing in shear. [*From M. Wakabayashi, T. Nakamura, and S. Morino, An experiment of steel reinforced concrete cruciform frames,* Bull. Disaster Prevention Res. Inst., *Kyoto University,* **23,** *75–110 (1973).*]

diaphragm is needed even though local stress is somewhat relieved by the presence of the concrete. Diaphragms are placed outside the tubes as shown in Fig. 3-97, so that the tubes can be filled with concrete. For the setup in Fig. 3-97*a*, cracks often begin at the corner of the stiffener owing to stress concentration, and thus the stiffener shapes shown in Fig. 3-97*b* and *c* are recommended (Kurobane, Hisamitsu, and Sakamoto, 1967).

3.5.6.3 Prefabricated Composite Member-to-Member Connection   Composite members are often prefabricated and connected at the construction site in the same way as precast-concrete structures as described in Sec. 4.6.3. Most composite structures in Europe are of this type, and prefabricated encased structural systems are often employed for multistory apartment houses in Japan.

Prefabricated composite structural systems are divided into frame systems and panel systems as in the case of precast-concrete structures;

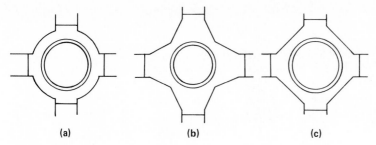

Figure 3-97 Stiffeners in the connection of beams to tube columns.

the former are further subdivided into linear systems and frame subassemblage systems. In any system connections must possess sufficient strength and ductility, and monolithic action of the connection should be ensured until the frame reaches the overall-failure state. Details of connections are similar to those for precast-concrete structures discussed in Sec. 4.6.3.2, although connections for prefabricated composite structures are more reliable, since steel sections encased in concrete can be directly connected by welding or fasteners.

3.5.7 Systems

As shown in Fig. 3-98, because of strain hardening a composite portal frame subjected to horizontal force with zero vertical load exhibits a gradual increase in strength after initial yielding. When $P/P_u = 0.4$, a negative slope appears in the horizontal-load–rotation relation because of the combined effects of the P-Δ moment and the deterioration of cross-sectional strength caused by the high axial force (Wakabayashi, 1974). Cyclic behavior in Fig. 3-99 shows a reduction in deformation capacity

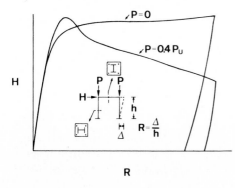

Figure 3-98 Load-deflection relationships of composite portal frames under constant vertical load and monotonic horizontal load. [*From M. Wakabayashi, Steel reinforced concrete, elastic plastic behavior of members, connections and frames*, Proc. Nat. Conf. Planning Des. Tall Build., *Tokyo, 3, 23–36 (1974)*.]

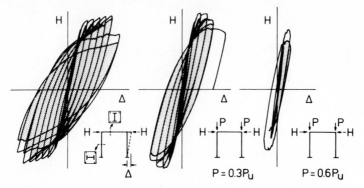

Figure 3-99 Load-deflection relationships for composite portal frames under constant vertical load and alternately repeated horizontal load. [*From M. Wakabayashi and K. Minami, Experimental studies on hysteretic characteristics of steel reinforced concrete columns and frames*, Proc. Int. Symp. Earthquake Struct. Eng., *University of Missouri—Rolla, St. Louis, 1, 467−480 (1976)*.]

with increasing axial force, similar to the behavior of encased beam columns in Fig. 3-89 (Wakabayashi and Minami, 1976).

3.5.8 Earthquake Damage

About 300 high-rise composite steel-and-concrete buildings of more than seven stories were subjected to the Miyagiken-Oki earthquake (1978). This was the first time that modern composite structures had experienced a severe earthquake. Structural skeletons such as beams and columns were not damaged even though accelerations of 100 percent of *g* were recorded in the upper stories. However, cracks were observed in many nonstructural reinforced-concrete exterior walls of high-rise apartment houses, and this raised the question of the design of nonstructural walls. An example is given in Fig. 3-100, where many cracks are observed in nonstructural walls with openings, while the columns and beams marked by dashed lines are not damaged at all (AIJ, 1980).

3.6 Behavior of Masonry Structures

3.6.1 Introduction

Masonry construction was the leading structural system for medium- and high-rise buildings until the beginning of the twentieth century, when it was replaced by new materials, such as reinforced-concrete and steel.

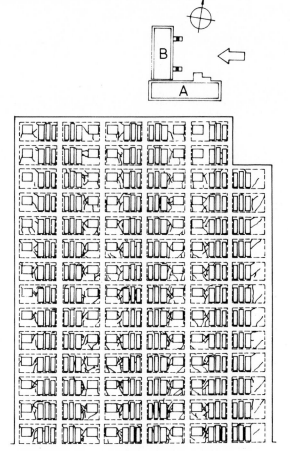

Figure 3-100   Crack pattern in nonstructural walls of an
apartment house, Miyagiken-Oki earthquake, Japan, 1978.
[*From AIJ*, Report on the Damage Due to 1978 Miyagiken-
Oki Earthquake, *Tokyo, 1980 (in Japanese).*]

However, it is still used in earthquake-prone areas because it is easy to
procure, economical, and good for insulation and finishing.

In the San Francisco (1906), Kanto (1923), and Hawke's Bay (1931)
earthquakes, many masonry structures suffered damage, and the inferi-
ority of the conventional masonry building as an earthquake-resistant
structure was revealed. Since then masonry structures have not been
constructed in countries such as Japan, where codes require that the
height of an unreinforced-masonry structure should be less than 9 m and
that of reinforced-masonry structures less than three stories. However, as

mentioned above, masonry structures are still constructed in many earthquake-prone areas, and many lives have been lost not only because of the failure of individual houses made of stone or adobe but also because of the failure of apartment houses, schools, and hospitals.

The reasons for poor performance of masonry structures in earthquakes are as follows:

1. The material itself is brittle, and strength degradation due to load repetition is severe.

2. Heavy weight.

3. Large stiffness, which leads to large response to earthquake waves of short natural period.

4. Large variability in strength depending on the quality of construction.

Nevertheless, masonry structures which were designed and constructed by taking these factors into consideration suffered little damage in the Alaska (1964) and San Fernando (1971) earthquakes, and research into earthquake-resistant masonry structures has been initiated. Experimental investigations of the static and dynamic behavior of masonry structures as a whole are urgently required.

The load-deformation relation for a masonry member failing in flexure is approximately of the elastoplastic type, but the ductility of a masonry member failing in shear is very small. The behavior of the whole system is therefore brittle. As a result, the Uniform Building Code (UBC) categorizes masonry structures as "box systems" and requires that they be designed against earthquake forces with a response factor equal to 1.33, which is twice that for moment-resisting frames. The same approach is taken in ATC-3.

Unreinforced masonry should not be used for large-size building structures in areas prone to severe eathquakes because of its brittle behavior. Only masonry structures which are reinforced are allowed in areas above Zone II in the United States. The following discussions deal primarily with reinforced masonry.

3.6.2 Types of Construction

Materials used in masonry structures vary widely from nonengineering materials such as stone and adobe to earthquake-resistant materials such as bricks and concrete blocks. The form of construction also varies from stone masonry without the use of mortar to reinforced masonry.

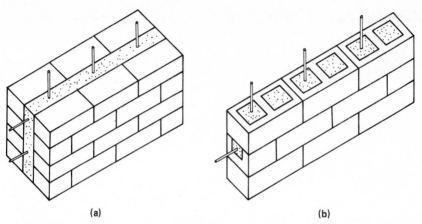

Figure 3-101 Typical forms of reinforced-masonry construction. *(a)* Reinforced grouted masonry. *(b)* Reinforced hollow-unit masonry.

Typical examples of reinforced masonry are shown in Fig. 3-101. In *a* reinforcing bars are placed between two layers of masonry, and the space is grouted with concrete which contains fine gravel as aggregate. In *b* hollow units are laid one by one, vertical and horizontal reinforcement is included, and mortar is in the holes. Mortar should be placed in vertical holes even if vertical reinforcement is not present.

3.6.3   Behavior of Materials

Compressive strengths of masonry elements are as follows: 20 to 140 MPa for solid clay bricks, 7 to 100 MPa for sand-lime building bricks, and 3.5 to 48 MPa for hollow concrete blocks (gross section). These values are quite high. On the other hand, the strength of mortar is lower: 0.1 to 1 MPa for lime mortar and 15 to 30 MPa for cement or cement-lime mortar. The compressive strength of a prism made by laying these materials lies between the strengths of the masonry element and the mortar. The prism strength is less than half the strength of the element; it increases with increasing mortar strength and with decreasing mortar thickness.

The following explanation of why prism strength depends on mortar strength refers to Fig. 3-102. The Poisson ratio for the joint mortar gradually increases as compressive stress approaches maximum strength. This develops tensile stress in the masonry element and eventually causes cracking (Priestley, 1980). A stainless-steel plate sandwiched in the joint could thus be used to restrain the transverse elongation of the joint mortar and increase the ductility of the prism (Priestley and Elder, 1982).

Figure 3-102 Failure
mechanism of masonry
prisms. [*From M. J. N. Priestley, Masonry, in E. Rosenblueth (ed.)*, Design of Earthquake Resistant Structures,
*Pentech Press, London, 1980,
pp. 195–222.*]

3.6.4 Members Failing in Flexure

3.6.4.1 In-Plane Bending of Walls   Flexural failure takes place in masonry
walls in which the height-to-length ratio is relatively large and the vertical
reinforcement content is not large. At failure, the extreme-compression
fiber strain reaches about 0.3 percent as in the case of reinforced-concrete
members, and the tension reinforcement yields. The ultimate flexural
strength of such a masonry wall can thus be calculated by Eq. (3-29), which
is an approximate strength formula for a reinforced-concrete wall.

The hysteretic behavior of a masonry wall failing in flexure under
repeated in-plane bending with low axial force is approximately of the
elastoplastic type and shows high ductility and little strength degradation.
A masonry wall failing in flexure and subjected to high axial force is not
necessarily ductile, and degradation is severe (Meli, 1974; Priestley and
Elder, 1982).

3.6.4.2 Out-of-Plane Bending of Walls   The behavior of a reinforced-
masonry wall subjected to out-of-plane bending is quite similar to that of a
reinforced-concrete wall, and ductility is very large. Ultimate strength of
such a masonry wall can be calculated by the formula for a reinforced-
concrete beam.

3.6.5 Members Failing in Shear

3.6.5.1 Walls, Piers, and Beams   In masonry walls without openings, shear
failure often takes place as shown in Fig. 3-103*a*, while in piers and beams
between openings failure is as in Fig. 3-103*b*. Since relative horizontal

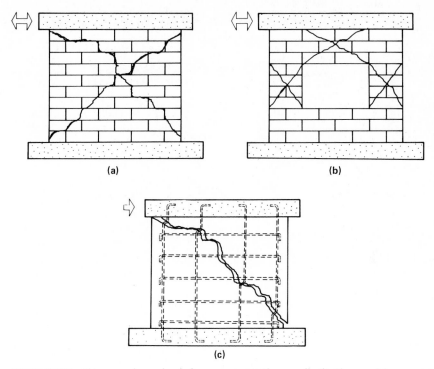

Figure 3-103 Shear cracks and reinforcement. *(a)* Shear wall. *(b)* Piers and beams. *(c)* Effectiveness of reinforcement.

movement occurs between the segments on either side of a crack, as shown in Fig. 3-103*c*, horizontal reinforcement is effective. On the other hand, little resistance is provided by dowel action of the vertical reinforcement. Therefore, horizontal reinforcement in the wall or in the pier and vertical reinforcement in the beam act effectively to resist shear.

Shear failure is more likely to occur in such wall elements as the height-to-length ratio becomes smaller, as the vertical reinforcement content, which mainly resists flexure, becomes larger, and as the horizontal reinforcement content becomes smaller. Shear failure tends to be brittle, with low energy-dissipation capacity, and severe strength degradation due to load repetition takes places (Mayes, Omote, and Clough, 1976, 1977; Sheppard, Terčelj, and Türnšek, 1977).

Figure 3-104 shows hysteresis loops for a pier subjected to repeated shear force (Wakabayashi and Nakamura, 1983). Ductility is improved by increasing shear reinforcement. The ductility of the pier is further improved by inserting confining plates into the joints which are subjected to high compressive stress as a result of flexure. The action of these plates

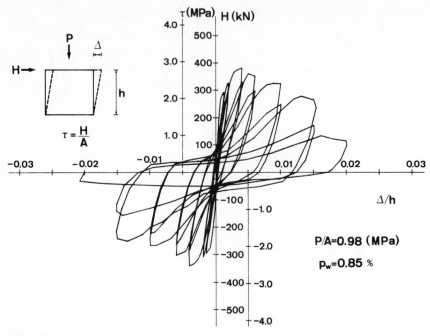

Figure 3-104 Horizontal load-deflection relationships for a pier subjected to reversed shear and failing in shear. [*From M. Wakabayashi and T. Nakamura, Experimental study on masonry walls, Proc. Kinki Branch Arch. Inst. Japan Ann. Conv., 461- 464 (1983; in Japanese).*]

was described in Sec. 3.6.3. The ductility of the wall is of course related to the ductility of the masonry materials (Priestley and Bridgeman, 1974; Priestley and Elder, 1982).

It has been reported that when repeated shear is applied to a masonry wall with a speed ranging from 1 to 5 cycles/s, strength increases in proportion to loading speed, the strength increment being about 15 percent of the static strength when loading speed is equal to 5 cycles/s (Terčelj, Sheppard, and Türnšek, 1977). Another report concludes that dynamic repeated loading causes severe capacity degradation and stiffness degradation even in a masonry wall in which flexure is predominant (Williams and Scrivener, 1974).

3.6.5.2 Infilled Walls   When a masonry wall surrounded by a reinforced-concrete frame is subjected to shear, the wall panel and the frame separate at a load equal to 50 to 70 percent of maximum capacity, and the wall then acts as a diagonal compression strut. Final failure occurs in the following ways (Priestley, 1980; Leuchars and Scrivener, 1976; Esteva, 1966):

1. *Failure of the diagonal compression strut.* Maximum capacity is determined by the maximum compression strength of the strut, which has a width of about one-fourth of the diagonal length of the panel (Fig. 3-105*a*).

2. *Horizontal sliding failure of the panel.* If resistance to sliding is smaller than strength as limited by the diagonal strut, the wall panel fails by horizontal sliding as shown in Fig. 3-105*b*. Once sliding takes place, external shear is resisted only by the columns since the friction at the sliding surfaces is very small. The resistance provided by the columns is determined as the smaller value of the sum of the column shear strengths and the sum of $4M_u/h$ for each column, which is the horizontal force forming plastic hinges at the center and the end of each column as shown in Fig. 3-105*b*. M_u designates the ultimate moment of the column section.

Infilled walls as described above show larger ductility than isolated masonry walls or piers. Strength deterioration is more severe for hollow masonry than for solid masonry because of crushing and spalling of the shell (Meli, 1974). The strength and energy-dissipation capacity of a frame with infilled walls are much higher than those of a bare frame, and thus a frame with infilled walls remains effective against earthquakes even though input forces increase because of the higher stiffness (Klinger and Bertero, 1977).

3.6.5.3 Nonstructural Infilled Walls   As discussed in Secs. 4.7.4 and 4.7.5, a nonstructural wall not separated from other structural elements increases stiffness and hence brings about a higher earthquake response. Such a wall causes stress concentration in particular members and/or

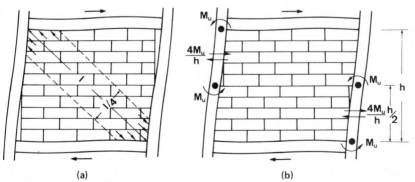

(a)   (b)

Figure 3-105 Failure and resisting mechanism of infilled walls. (*a*) Failure of a diagonal strut. (*b*) Horizontal sliding of a panel.

torsional deformation of the frame. On the other hand, it increases story shear strength and energy-dissipation capacity.

3.6.6 Behavior of Systems

Few investigations have been conducted into the deformation behavior of masonry systems. Analysis of masonry systems is difficult, since they are designed to minimize openings. An approximate solution may be obtained by elastic or elastoplastic analysis of a frame consisting of linear members which replaces the real structure with infilled walls. In the analysis, the effects of shear deformations and rigid zones around the beam-to-column connections can be taken into account if necessary. Shear failure often occurs in piers and beams of a masonry system as shown in Fig. 3-106, since the length-to-width ratio of these members is small. Once the boundary beams fail, following shear failure of the spandrel beams shown in Fig. 3-106*b*, the two walls resist external shear by cantilever action.

Besides failure due to in-plane earthquake forces, the failure patterns shown in Fig. 3-107 are often observed in masonry structures:

1. Flexural failure due to out-of-plane earthquake forces

2. Failures of the gable wall, the roof structure, and the intersection of the wall and the ridge

Because of these failures, the wall spalls off as a whole, thus endangering the surrounding area.

3.6.7 Earthquake Damage

Too many masonry structures have been damaged by past earthquakes. For example, among 50,000 building structures subjected to the Skopje earthquake (1963), 10 percent completely collapsed and 37 percent were seriously damaged and demolished (Kunze, Fintel, and Amrhein, 1965).

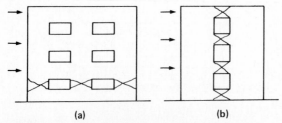

(a)   (b)

Figure 3-106 Diagonal cracks in piers and beams, *(a)* Cracks in piers. *(b)* Cracks in beams.

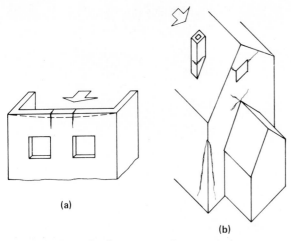

(a)

(b)

Figure 3-107 Miscellaneous crack patterns in masonry buildings. *(a)* Cracks due to out-of-plane bending. *(b)* Cracks due to local stress and other stress.

Most of these structures were four- to five-story masonry or mixed masonry and reinforced-concrete structures in which adequate design attention had not been given to severe earthquake action. Figure 3-108 shows an example of damage to a masonry building. Figure 3-109 is an example of damage to an unreinforced-masonry school building.

Figure 3-108 Damage to an unreinforced-brick-masonry house, Montenegro earthquake, Yugoslavia, 1979. *(Courtesy of B. Simeonov.)*

Figure 3-109 Damage to an unreinforced-masonry school building, Tang-shan earthquake, China, 1976.

3.7 Behavior of Timber Structures

3.7.1 Introduction

Timber is light and possesses high strength and ductility. Thus, it is counted as a good earthquake-resistant material. Timber structures can be considered to possess high earthquake-resistant capacity if slabs and roofs are light. The number of lives lost due to the collapse of timber structures is probably much smaller than that due to the collapse of masonry structures.

However, it must be noted that a substantial number of people died or were injured as a result of the fire after the San Francisco earthquake (1906). Similarly, most of the 140,000 lives lost in the Kanto earthquake (1923) were due to the fire which occurred after the earthquake in the area of dense timber housing. Fire caused by earthquakes in timber houses in urban areas is a very important problem.

Timber structures are divided into light construction, such as houses, which requires no structural calculation, and engineering construction, for which earthquake-resistant capacity must be checked by calculation. In both cases, timber structures are usually composed of diaphragms and shear walls.

3.7.2 Shear Walls

Various materials are used for shear walls and diaphragms in timber structures: panels sheathed with plywood, wood-stud walls sheathed with lath and plaster, gypsum sheathing board, gypsum wallboard, and fiberboard sheathing. Allowable strengths for these materials, as determined by shear tests, are given in ATC-3 and other codes, but it should be noted that specified strengths differ from country to country.

Figure 3-110 shows load-deflection relationships for timber shear walls with various sheathings, from which it is observed that strength varies substantially with the type of sheathing. The relationships are nonlinear from the early stage of loading, and ductility is large (Watabe and Kawashima, 1971). The behavior of a timber wall subjected to repeated shear is shown in Fig. 3-111 (Medearis, 1966). Most timber walls possess a large equivalent damping ratio equal to 10 percent regardless of deflection amplitude.

3.7.3 Systems

The hysteretic behavior of a timber system is affected by the strength and deformation behavior of the anchorages of the shear wall and also of the diaphragm. However, the behavior of the system is in general quite similar to that of the components, i.e., the shear wall and the diaphragm.

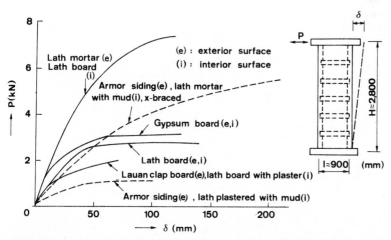

Figure 3-110 Load-deflection relationships for timber walls subjected to monotonic shear force.

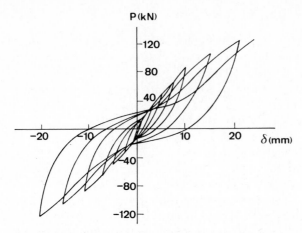

Figure 3-111   Load-deflection relationship for a timber wall under alternately repeated shear force. [*From K. Medearis, Static and dynamic properties of shear structures,* Proc. Int. Symp. Effect Repeated Loading Mater. Struct., *RILEM, Mexico City, 6 (1966).*]

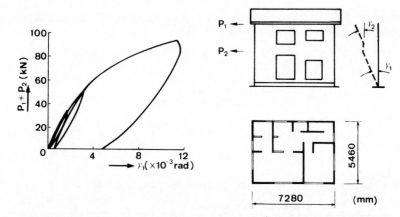

Figure 3-112   Horizontal load-deflection relationship for a timber house. [*From H. Noguchi and H. Sugiyama, On the racking stiffness of bearing walls of platform construction due to full-scale house test and ordinary racking test (no. 1),* Trans. Arch. Inst. Japan, *no. 248, 1–12 (1976; in Japanese).*]

Figure 3-112 shows hysteresis loops for a two-story timber house in which torsion is not observed since the walls are placed symmetrically as a whole. Eccentric placement of walls causes torsion.

3.7.4 Earthquake Damage

Although the earthquake-resistant capacity of timber structures is quite high, many nonengineering timber structures have been badly damaged

Figure 3-113 Damage to timber split-level houses, San Fernando earthquake, California, 1971. (*From G. A. Frazier, J. H. Wood, and G. W. Housner,* Earthquake Damage to Buildings; Engineering Features of the San Fernando Earthquake, *EERL 71-02, Earthquake Eng. Res. Lab., California Institute of Technology, Pasadena, Calif., 1971, pp. 140−298.*)

by past earthquakes because of inadequate structural shape and details. About 7000 timber houses completely collapsed or were seriously damaged by the Miyagiken-Oki earthquake of 1978 (Iizuka, 1980). The main causes of the damage are as follows:

1. Insufficient number of walls
2. Torsion due to eccentric placement of walls
3. Overweight of roofs
4. Poor anchorage either of a column to a sill or of a sill to a foundation
5. Separation of foundations
6. Poor connections
7. Landslide and other soil failures

Damage to timber structures was centered on an area of thick alluvium, as described in Sec. 1.3.3 (Ohsaki, 1962; SRT, 1960). Figure 3-113 shows the collapse of a garage in the first story, which was typical of the damage observed in the San Fernando earthquake of 1971 (Frazier, Wood, and Housner, 1971).

REFERENCES

ACI Committee 318 (1983a). *Building Code Requirements for Reinforced Concrete (ACI 318-83)*, American Concrete Institute, Detroit.

———— (1983b). *Commentary on Building Code Requirements for Reinforced Concrete (ACI 318-83)*, American Concrete Institute, Detroit.

AIJ (Architectural Institute of Japan) (1968). *Report on the Damage Due to 1968 Tokachi-Oki Earthquake*, AIJ, Tokyo (in Japanese).

———— (1975a). *Guide of the Plastic Design of Steel Structures*, AIJ, Tokyo (in Japanese).

———— (1975b). *Design Standard for Composite Steel-Reinforced Concrete Structures*, AIJ, Tokyo (in Japanese).

———— (1979). *Design Standard for Steel Structures*, AIJ, Tokyo.

———— (1980). *Report on the Damage Due to 1978 Miyagiken-Oki Earthquake*, AIJ, Tokyo (in Japanese).

AISC (American Institute of Steel Construction) (1969). *AISC Specification for the Design, Fabrication and Erection of Structural Steel for Buildings, and Commentary*, AISC, New York.

———— (1978). *AISC Specification for the Design, Fabrication and Erection of Structural Steel for Buildings, and Commentary*, AISC, New York.

Aoyama, H. (1981). Mechanical properties of steel and concrete under cyclic loading, in O. Ergünay and M. Erdik (eds.), *State-of-the-Art in Earthquake Engineering*, Ankara, 301−322.

————, S. Sugano, and S. Nakata (1970). Vibration and static loading tests of Hachinohe Technical College, part 2: Static loading tests, *Trans. Arch. Inst. Japan*, no. 169, 33−41 (in Japanese).

Arnold, P., and D. F. Adams (1968). Strength and behavior of an inelastic hybrid frame, *J. Struct. Div., Am. Soc. Civ. Eng.*, **94**(ST-1), 243−266.

ASCE-WRC (Joint Committee of the Welding Research Council and the American Society of Civil Engineers) (1971). *Plastic Design in Steel: A Guide and Commentary*, 2d ed., New York.

Astaneh-Asl, A., S. C. Goel, and R. D. Hanson (1982). *Cyclic Behavior of Double Angle Bracing Members with End Gusset Plates*, UMEE 82R7, Dept. Civ. Eng., University of Michigan, Ann Arbor.

ATC-3 (Applied Technology Council) (1978). *Tentative Provisions for the Development of Seismic Regulations for Buildings (ATC-3-06)*, Nat. Bur. Stand. Spec. Publ. 510, Washington.

Bertero, V. V. (1969). Seismic behavior of steel beam-to-column connection subassemblages, *Proc. Fourth World Conf. Earthquake Eng.*, Santiago, **2**(B-3), 31−44.

————, H. Krawinkler, and E. P. Popov (1973). *Further Studies on Seismic Behavior of Steel Beam-to-Column Subassemblages*, Rep. 73-27, Earthquake Eng. Res. Center, College of Engineering. University of California, Berkeley.

———— and E. P. Popov (1965). Effect of large alternating strains of steel beams, *J. Struct. Div., Am. Soc. Civ. Eng.*, **91**(ST-1), 1−12.

————, ————, and H. Krawinkler (1972). Beam-to-column subassemblages under repeated loading, *J. Struct. Div., Am. Soc. Civ. Eng.*, **98**(ST-5), 1137−1159.

Black, R. G., W. A. Wenger, and E. P. Popov (1980). *Inelastic Buckling of Steel Struts under Cyclic Load Reversals*, Rep. UCB/EERC-80/40, Earthquake Eng. Res. Center, College of Engineering, University of California, Berkeley.

Blakeley, R. W. G., and R. Park (1973). Prestressed concrete sections with cyclic flexure, *J. Struct. Div., Am. Soc. Civ. Eng.*, **99**(ST-8), 1717−1742.

Bresler, B., and J. G. MacGregor (1967). Review of concrete beams failing in shear, *J. Struct. Div., Am. Soc. Civ. Eng.*, **93**(ST-1), 343−372.

Cardenas, A. E., J. M. Hanson, W. G. Corley, and E. Hognestad (1973). Design provisions for shear walls, *J. Am. Concr. Inst.*, **70**(3), 221−230.

Carpenter, L. D., and L.-W. Lu (1969). Repeated and reversed load tests on full-scale steel frames, *Fourth World Conf. Earthquake Eng.*, Santiago, **1**(B-2), 125−136.

Corley, W. G. (1966). Rotational capacity of reinforced concrete beams. *J. Struct. Div., Am. Soc. Civ. Eng.*, **92**(ST-5), 121−146.

———— and N. W. Hanson (1969). Design of beam column joints for seismic resistant reinforced concrete frames, *Fourth World Conf. Earthquake Eng.*, Santiago, **2**(B-3), 69–82.

Daniels, J. H., and J. W. Fisher (1970). Behavior of composite beam to column joints, *J. Struct. Div., Am. Soc. Civ. Eng.*, **96**(ST-3), 671–685.

———— and L.-W. Lu (1966). *The Subassemblage Method of Designing Unbraced Multi-Story Frames*, Fritz Eng. Lab. Rep. 273.37, Lehigh University, Bethlehem, Pa.

de Buen, O. (1969). A modification to the subassemblage method of designing unbraced multistory frames, *Eng. J.*, American Institute of Steel Construction, **6**(4).

———— (1980). Steel structures, in E. Rosenblueth (ed.), *Design of Earthquake Resistant Structures*, Pentech Press, London, 100–141.

Driscoll, G. C., Jr., L. S. Beedle, T. V. Galambos, L.-W. Lu, J. W. Fisher, A. Ostapenko, J. Daniels, and B. Parikh (1965). *Plastic Design of Multi-Story Frames—Lecture Notes and Design Aids,* Fritz Eng. Lab. Rep. 273.20 and 273.24, Lehigh University, Bethlehem, Pa.

Esteva, L. (1966). Behavior under alternating loads of masonry diaphragms framed by reinforced concrete members. *Proc. Int. Symp. Effect Repeated Loading Mater. Struct.*, RILEM, Mexico City, **5.**

Fenwik, R. C., and H. M. Irvine (1977). Reinforced concrete beam-column joints for seismic loading, *Bull. N.Z. Nat Soc. Earthquake Eng.*, **10**(3), 121–128.

Fiorato, A. E., R. G. Oesterle, Jr., and J. E. Carpenter (1976). Reversing load tests of five isolated structural walls, *Proc. Int. Symp. Earthquake Struct. Eng.*, University of Missouri–Rolla, St. Louis, 437–453.

Frazier, G. A., J. H. Wood, and G. W. Housner (1971). *Earthquake Damage to Buildings; Engineering Features of the San Fernando Earthquake*, EERL 71-02, Earthquake Eng. Res. Lab., California Institute of Technology, Pasadena, 140–298.

Fujimoto, M., T. Aoyagi, K. Ukai, A. Wada, and K. Saito (1972). Structural characteristics of eccentric K-braced frames, *Trans. Arch. Inst. Japan*, no. 195, 39–49 (in Japanese).

————, A. Wada, K. Shirakata, and R. Kosugi (1973). Nonlinear analysis for K-type braced steel frames, *Trans. Arch. Inst. Japan*, no. 209, 41–51 (in Japanese).

Fujiwara, T. (1980). Seismic behavior of inelastic members of braced frame structure, *Proc. Seventh World Conf. Earthquake Eng.*, Istanbul, **7,** 241–248.

Galambos, T. V., and M. G. Lay (1965). Studies of the ductility of steel structures, *J. Struct. Div., Am. Soc. Civ. Eng.*, **91**(ST-4), 125–151.

Gavrilovic, P., M. Velkov, D. Jurukovski, and D. Mamucevski (1980). Behavior of reinforced concrete beam-column joints under cyclic loading, *Proc. Seventh World Conf. Earthquake Eng.*, Istanbul, **7,** 289–296.

Gugerli, H., and S. C. Goel (1982). *Inelastic Cyclic Behavior of Steel Bracing Members*, UMEE 82R1, Dept. Civ. Eng., University of Michigan, Ann Arbor.

Hanson, R. (1975). Characteristics of steel members and connections, *Proc. U.S. Nat. Conf. Earthquake Eng.*, Ann Arbor, Mich., 255–267.

Hognestad, E. (1952). Inelastic behavior in tests of eccentrically loaded short reinforced concrete columns, *J. Am. Concr. Inst.* **24**(2).

Igarashi, S., K. Inoue, M. Asano, and K. Ogawa (1972). Hysteretic characteristics of steel braced frames, part 2: Dynamic analysis of diagonal bracings, *Trans. Arch. Inst. Japan*, no. 205, 37–42 (in Japanese).

Iizuka, G. (1980). On the damage and caution to wooden houses by recent earthquakes in Japan, *Proc. Seventh World Conf. Earthquake Eng.*, Istanbul, **5,** 121–124.

Inomata, S. (1979). Background of the FIP specification for aseismic design of prestressed concrete structures, *Prestressed Concr.*, **21**(4), 62–71 (in Japanese).

Jain, A. K., S. C. Goel, and R. D. Hanson (1980). Hysteretic cycles of axially loaded steel members, *J. Struct. Div., Am. Soc. Civ. Eng.*, **106**(ST-8), 1777–1795.

———— and R. D. Hanson (1980). Hysteresis models of steel members for earthquake response of braced frames, *Proc. Seventh World Conf. Earthquake Eng.*, Istanbul, **6,** 463–470.

Jirsa, J. O. (1981). Seismic behavior of R.C. connections (beam-column joints), in O. Ergünay and M. Erdik (eds.), *State-of-the-Art in Earthquake Engineering 1981*, Ankara, 365–374.

————, D. F. Meinheit, and J. W. Woolen (1975). Factors influencing the shear strength of beam-column joints, *Proc. U.S. Nat. Conf. Earthquake Eng.*, Ann Arbor, Mich., 297–305.

Kamimura, T. (1975). Ultimate shear capacity of reinforced concrete beam-to-column connection, *Proc. Arch. Inst. Japan Ann. Conv.*, 1155–1156.

Kato, B. (1982). Beam-to-column connection research in Japan, *J. Struct. Div., Am. Soc. Civ. Eng.*, **108**(ST-2), 343–360.

—— and H. Akiyama (1973). Theoretical prediction of the load-deflection relationship of steel members and frames, *Symposium on Resistance and Ultimate Deformability of Structures Acted on by Well-Defined Repeated Loads: Preliminary Report*, International Association of Bridge and Structural Engineering, Lisbon Symposium, 23–28.

—— and M. Nakao (1973). The influence of the elastic plastic deformation of beam-to-column connections on the stiffness, ductility and strength of open frames, *Proc. Fifth World Conf. Earthquake Eng.*, Rome, **1**, 825–828.

—— and N. Uchida (1974). Test of composite beam to column joints, *Proc. Nat. Conf. Planning Des. Tall Build.*, Tokyo, **3**, 119–120.

Kent, D. C., and R. Park (1971). Flexural members with confined concrete, *J. Struct. Div., Am. Soc. Civ. Eng.*, **97**(ST-7), 1969–1990.

Klinger, R. E., and V. V. Bertero (1977). Infilled frames in aseismic construction, *Proc. Sixth World Conf. Earthquake Eng.*, New Delhi, **2**, 1926–1932.

Krawinkler, H., V. V. Bertero, and E. P. Popov (1971). *Inelastic Behavior of Steel Beam-to-Column Subassemblages*, Rep. EERC 71-7, Earthquake Eng. Res. Center, College of Engineering, University of California, Berkeley.

——, ——, and —— (1975). Shear behavior of steel frame joints, *J. Struct. Div., Am. Soc. Civ. Eng.*, **101**(ST-11), 2317–2336.

—— and E. P. Popov (1982). Seismic behavior of moment connections and joints, *J. Struct. Div., Am. Soc. Civ. Eng.*, **108**(ST-2), 373–391.

Kunze, W. E., M. Fintel, and J. E. Amrhein (1965). The Skopje, Yugoslavia, Earthquake of July 26, 1963, *Proc. Third World Conf. Earthquake Eng.*, Auckland, **3**, S-39–48.

Kurobane, Y., S. Hisamitsu, and S. Sakamoto (1967). On the strength and stiffness of the corner connection in the steel pipe column, *Trans. Arch. Inst. Japan*, no. 132, 17–24 (in Japanese).

Lay, M. G., and T. V. Galambos (1965). Inelastic steel beams under uniform moment, *J. Struct, Div., Am. Soc. Civ. Eng.*, **91**(ST-6), 67–93.

—— and —— (1967). Inelastic beams under moment gradient, *J. Struct. Div., Am. Soc. Civ. Eng.*, **93**(ST-1), 381–399.

Leuchars, J. M., and J. C. Scrivener (1976). Masonry infill panels subjected to cyclic in-plane loading, *Bull. N. Z. Nat. Soc. Earthquake Eng.*, **9**(2), 121–131.

Maison, B. F., and E. Popov (1980). Cyclic response prediction for braced steel frames, *J. Struct. Div., Am. Soc. Civ. Eng.*, **106** (ST-7), 1401–1416.

Makino, M., M. Ozaki, and M. Hirosawa (1965). *Bearing Walls of Steel Framed Reinforced Concrete with Steel Plates*, Rep. Build. Res. Inst., no. 46, 99–130.

Mattock, A. H. (1967). Discussion of rotational capacity of reinforced concrete beams by W. G. Corley, *J. Struct. Div., Am. Soc. Civ. Eng.*, **93**(ST-2), 519–522.

Mayes, R. L., Y. Omote, and R. W. Clough (1976). *Cyclic Shear Tests of Masonry Piers*, Rep. EERC 76-8, Earthquake Eng. Res. Center, College of Engineering, University of California, Berkeley.

——, ——, and —— (1977). Cyclic shear tests on masonry piers, *Proc. Sixth World Conf. Earthquake Eng.*, New Delhi, **3**, 3195–3201.

Medearis, K. (1966). Static and dynamic properties of shear structures, *Proc. Int. Symp. Effect Repeated Loading Mater. Struct.*, RILEM, Mexico City, **6**.

Meli, R. (1974). Behavior of masonry walls under lateral loads, *Proc. Fifth World Conf. Earthquake Eng.*, Rome, **1**, 853–862.

Minami, K., and M. Wakabayashi (1974). Shear strength of steel reinforced concrete (SRC) columns, *Symposium on Design and Safety of Reinforced Concrete Compression Members: Preliminary Report*, International Association of Bridge and Structural Engineering, Quebec Symposium, 305–313.

—— and —— (1980). Earthquake resistant properties of diagonally reinforced concrete columns, *Trans. Japan Concr. Inst.*, **2**, 431–438.

——— and ——— (1981). Rational analysis of shear in reinforced concrete columns, IABSE Colloquium, Delft, 1981, *Advanced Mechanics of Reinforced Concrete: Final Report*, International Association of Bridge and Structural Engineering, Delft, Netherlands, 603–614.

Mitani, I., M. Makino, and C. Matsui (1977). Influence of local buckling on cyclic behavior of steel beam-columns, *Proc. Sixth World Conf. Earthquake Eng.*, New Delhi, **3**, 3175–3180.

Morita, S., and T. Kaku (1973). Local bond stress-slip relationship under repeated loading, *Symposium on Resistance and Ultimate Deformability of Structures Acted on by Well-Defined Repeated Loads: Preliminary Report*, International Association of Bridge and Structural Engineering, Lisbon Symposium, 221–227.

Muguruma, H., F. Watanabe, and E. Nagai (1978). A study on the hysteretic characteristics of prestressed concrete statically indeterminate rigid frame under repeated horizontal load, *Rep. Kinki Branch Arch. Inst. Japan*, 73–76 (in Japanese).

———, ———, H. Tanaka, K. Sakurai, E. Nakamura, and S. Shoda (1979). Mathematical model of compressive stress strain curve of laterally reinforced concrete, *Rep. Kinki Branch Arch. Inst. Japan*, no. 19, 17–20 (in Japanese).

Naka, T., B. Kato, M. Watabe, and M. Nakao (1969). Research on the behavior of steel beam-to-column connections in the seismic-resistant structure, *Proc. Fourth World Conf. Earthquake Eng.*, Santiago, 2(B-3), 1–14.

Nakamura, T., and M. Wakabayashi (1976). A study on the superposition method to estimate the ultimate strength of steel reinforced concrete column subjected to axial thrust and bending moment simultaneously, *Bull. Disaster Prevention Res. Inst.*, Kyoto University, **26**, 163–193.

Noguchi, H., and H. Sugiyama (1976). On the racking stiffness of bearing walls of platform construction due to full-scale house test and ordinary racking test (no. 1), *Trans. Arch. Inst. Japan*, no. 248, 1–12 (in Japanese).

Nonaka, T. (1973). An elastic-plastic analysis of a bar under repeated axial loading, *Int. J. Solids Struct.*, **9**, 569–580.

Ohsaki, Y. (1962). Earthquake damage of wooden buildings and depth of alluvial deposit, *Trans. Arch. Inst. Japan*, no. 72, 29–32.

Park, R. (1980). Partially prestressed concrete in seismic design of frames, *Proc. FIP Symp. Partial Prestrassing Practical Constr. Prestressed Reinf. Concr.*, Bucharest, **1**, 104–117.

——— and T. Paulay (1975). *Reinforced Concrete Structures*, John Wiley & Sons, Inc., New York.

——— and ——— (1980). Concrete structures, in E. Rosenblueth (ed.), *Design of Earthquake Resistant Structures*, Pentech Press, London, 142–194.

——— and K. J. Thompson (1977). Some recent research in New Zealand into aspects of the seismic resistance of prestressed concrete frames, *Proc. Sixth World Conf. Earthquake Eng.*, New Delhi, **3**, 3227–3232.

Paulay, T. (1972). Some aspects of shear wall design, *Bull. N.Z. Nat. Soc. Earthquake Eng.*, **5**(3), 89–105.

——— (1980). Earthquake-resisting shear walls—New Zealand design trends, *J. Am. Concr. Inst.*, **77**, 144–152.

———, R. Park, and M. J. N. Priestley (1978). Reinforced concrete beam-column joints under seismic actions, *J. Am. Concr. Inst.*, **75**(11), 585–593.

——— and A. Scarpas (1981). The behavior of exterior beam-column joints, *Bull. N.Z. Nat. Earthquake Eng.*, **14**(3), 131–144.

Pfrang, E. O., C. P. Siess, and M. A. Sozen (1964). Load-moment-curvature characteristics of reinforced concrete cross sections, *J. Am. Concr. Inst.*, **61**(7), 763–778.

Popov, E. P. (1980a). An update on eccentric seismic bracing, *Eng. J.*, American Institute of Steel Construction, third quarter, 70–71.

——— (1980b). Eccentric seismic bracing of steel frames, *Proc. Seventh World Conf. Earthquake Eng.*, Istanbul, **7**, 127–132.

——— and V. V. Bertero (1980). Seismic analysis of some steel building frames, *J. Eng. Mech. Div., Am. Soc. Civ. Eng.*, **106**, 75–93.

——, ——, and S. Chandramouli (1975). Hysteretic behavior of steel columns, *Proc. U.S. Nat. Conf. Earthquake Eng.*, Ann Arbor, Mich., 245–254.

——, ——, and H. Krawinkler (1974). Moment-resisting steel subassemblages under seismic loadings, *Proc. Fifth World Conf. Earthquake Eng.*, Rome, **2**, 1481–1490.

—— and R. B. Pinkney (1968). Behavior of steel building connections subjected to inelastic strain reversals, *Steel Res. Constr. Bull.*, no. 13, American Iron and Steel Institute, New York.

—— and —— (1969a). Reliability of steel beam-to-column connections under cyclic loading, *Proc. Fourth World Conf. Earthquake Eng.*, Santiago, **2**(B-3), 15–30.

—— and —— (1969b). Cyclic yielding reversal in steel building connections, *J. Struct. Div., Am. Soc. Civ. Eng.*, **95**(ST-3), 327–353.

—— and R. M. Stephan (1972). Cyclic loading of full-size steel connections, *Steel Res. Constr., Bull.* no. 21, American Iron and Steel Institute, New York.

——, K. Takanashi, and C. Roeder (1976). *Structural Steel Bracing Systems:Behavior under Cyclic Loading*, Rep. EERC/76-17, Earthquake Eng. Res. Center, College of Engineering, University of California, Berkeley.

——, V. Zayas, and S. Mahin (1979). Cyclic inelastic buckling of thin tubular columns, *J. Struct. Div., Am. Soc. Civ. Eng.*, **105**(ST-11), 2261–2277.

Priestley, M. J. N (1980). Masonry, in E. Rosenblueth (ed.), *Design of Earthquake Resistant Structures*, Pentech Press, London, 195–222.

—— and D. O. Bridgeman (1974). Seismic resistance of brick masonry walls, *Bull. N.Z. Nat. Soc. Earthquake Eng.*, **7**(4), 167–187.

—— and D. McG. Elder (1982). Cyclic loading tests of slender concrete masonry shear walls, *Bull. N.Z. Nat. Soc. Earthquake Eng.*, **15**(1), 3–21.

Roeder, C., and E. P. Popov (1978). Eccentrically braced steel frames for earthquakes, *J. Struct. Div., Am. Soc. Civ. Eng.*, **106**(ST-3), 391–413.

Sakamoto, J., and Y. Kohama (1973). On the dynamic behavior of braced structures subjected to random excitation, *Trans. Arch. Inst. Japan*, no. 248, 31–42 (in Japanese).

—— and A. Miyamura (1966). Critical strength of elastoplastic steel frames under vertical and horizontal loadings, *Trans. Arch. Inst. Japan*, no. 124, 1–7 (in Japanese).

Scott, B. D., R. Park, and M. J. N. Priestley (1982). Stress-strain behavior of concrete confined by overlapping hoops at low and high strain rates, *J. Am. Concr. Inst.*, **79**(2), 13–27.

Sheikh, S. A. (1982). A comparative study of confinement models, *J. Am. Concr. Inst.*, **79**(4), 296–306.

—— and S. M. Uzumeri (1980). Mechanism of confinement in tied columns, *Proc. Seventh World Conf. Earthquake Eng.*, Istanbul, **7**, 71–78.

Sheppard, P., S. Terčelj, and V. Türnšek (1977). The influence of horizontally-placed reinforcement on the shear strength and capacity of masonry walls, *Proc. Sixth World Conf. Earthquake Eng.*, New Delhi, **3**, 3000–3005.

Shibata, M., T. Nakamura, N. Yoshida, S. Morino, T. Nonaka, and M. Wakabayashi (1974). Elastic-plastic behavior of steel braces under repeated axial loading, *Proc. Fifth World Conf. Earthquake Eng.*, Rome, **1**, 845–848.

SRT (Subsoil Research Team of the Earthquake Research Institute) (1960). Earthquake damage and subsoil conditions as observed in certain districts of Japan, *Proc. Second World Conf. Earthquake Eng.*, Tokyo, **1**, 311–325.

Suzuki, T., and T. Ono (1977). An experimental study on inelastic behavior of steel members subjected to repeated loading, *Proc. Sixth World Conf. Earthquake Eng.*, New Delhi, **3**, 3163–3168.

Takanashi, K. (1974). Inelastic lateral buckling of steel beams subjected to repeated and reversed loadings, *Proc. Fifth World Conf. Earthquake Eng.*, Rome, **1**, 795–798.

Takeda, J. (1971). New steel bracings, *Kenchiku-Gijutsu*, no. 250, 121–128 (in Japanese).

Tall Building Committee 15 (Committee 15 of the Council on Tall Building and Urban Habitat) (1979). Plastic analysis and design, in *Monograph on Planning and Design of Tall Buildings*, SB-3, American Society of Civil Engineers, New York, 137–238.

Tall Building Committee 21D (Committee 21D of the Council on Tall Building and Urban Habitat) (1979). Design of cast-in-place concrete, in *Monograph on Planning and Design of Tall Buildings*, CB-11, American Society of Civil Engineers, New York, 501–576.

Tall Building Committee 21E (Committee 21E of the Council on Tall Building and Urban Habitat) (1979). Design of structures with precast concrete elements, in *Monograph on Planning and Design of Tall Buildings*, CB-12, American Society of Civil Engineers, New York, 575–653.

Tall Building Committee A41 (Committee A41 of the Council on Tall Building and Urban Habitat) (1979). Mixed construction, in *Monograph on Planning and Design of Tall Buildings*, SB-9, American Society of Civil Engineers, New York, 617–804.

Tall Building Committee 43 (Committee 43 of the Tall Building and Urban Habitat) (1979). Connections, in *Monograph on Planning and Design of Tall Buildings*, SB-7, American Society of Civil Engineers, New York, 483–575.

Tanabashi, R., K. Kaneta, and T. Ishida (1974). On the rigidity and ductility of steel bracing assemblage, *Proc. Fifth World Conf. Earthquake Eng.*, Rome, **1**, 834–840.

———, Y. Yokoo, M. Wakabayashi, T. Nakamura, and H. Kunieda (1971). Deformation history dependent inelastic stability of columns subjected to combined alternately loading, *RILEM Colloq. Int. Symp.*, Buenos Aires.

Terčelj, S., P. Sheppard, and V. Türnšek (1977). The influence on the shear strength and ductility of masonry walls in dynamic loading tests, *Proc. Sixth World Conf. Earthquake Eng.*, New Delhi, **3**, 2992–2997.

Tomii, M., C. Matsui, and K. Sakino (1974). Concrete filled steel tube structures, *Proc. Nat. Conf. Tall Build.*, Tokyo, **3**, 55–72.

Umemura, H., H. Aoyama, and H. Takizawa (1974). Analysis of the behavior of reinforced concrete structures during strong earthquakes based on empirical estimation of inelastic restoring force characteristics of members, *Proc. Fifth World Conf. Earthquake Eng.*, Rome, **2**, 2201–2210.

Uzumeri, S. M. (1977). Strength and ductility of cast-in-place beam-column joints, in *Reinforced Concrete Structures in Seismic Zones*, American Concrete Institute, SP53, 293–350.

Vann, W. P., L. E. Thompson, L. E. Whalley, and L. D. Ozier (1974). Cyclic behavior of rolled steel members, *Proc. Fifth World Conf. Earthquake Eng.*, Rome, **1**, 1187–1193.

Viest, I. M. (1974). Composite steel-concrete construction, *J. Struct. Div., Am. Soc. Civ. Eng.*, **100**(ST-5); 1085–1139.

Viwathanatepa, S., E. P. Popov, and V. V. Bertero (1979). *Effects of Generalized Loadings on Bond of Reinforcing Bars Embedded in Confined Concrete Blocks*, Rep. EERC-79-22, Earthquake Eng. Res. Center, College of Engineering, University of California, Berkeley.

Vogel, U. (1963). On the strength of steel frames. *Stahlbau*, **32**(4), 106–113.

Wakabayashi, M. (1965). The restoring force characteristics of multistory frames, *Bull. Disaster Prevention Res. Inst.*, Kyoto University, **14**, part 2, no. 78, 29–47.

——— (1970). The behavior of steel frames with diagonal bracings under repeated loading, *Proc. U.S.–Japan Semin. Earthquake Eng. with Emphasis on Safety School Build.*, Sendai, Japan, 328–345.

——— (1972). Frames under strong impulsive, wind or seismic loading, *Proc. Int. Conf. Planning Des. Tall Build.*, Lehigh University, Bethlehem, Pa., **2**(ST-15), 343–363.

——— (1973). Studies on damping and energy absorption of structures, *Symposium on Resistance and Ultimate Deformability of Structures Acted on by Well-Defined Repeated Loads: Introductory Report*, International Association of Bridge and Structural Engineering, Lisbon Symposium, 27–46.

——— (1974). Steel reinforced concrete, elastic plastic behavior of members, connections and frames, *Proc. Nat. Conf. Planning Des. Tall Build.*, Tokyo, **3**, 23–36.

——— (1976). Recent Japanese developments in mixed structures, *Proc. Am. Soc. Civ. Eng. Specialty Conf.*, **1**, 497–515.

——— (1977a). A new design method of long composite beam-columns, *Int. Colloq. Stability Struct. under Static and Dynamic Loads*, Washington, 742–756.

—— (1977*b*). Behavior of systems, panel 2: Behavior as related to design criteria, *Proc. Sixth World Conf. Earthquake Eng.*, New Delhi, **3**, 65–76.

—— (1982). Behavior of braces and braced frames under earthquake loadings, *Int. J. Struct.*, **2**(2), 49–70.

——, C. Matsui, K. Minami, and I. Mitani (1974*a*). Inelastic behavior of full scale steel frames with and without bracings, *Bull. Disaster Prevention Res. Inst.*, Kyoto University, **24**, part 1, no. 216, 1–23.

——, ——, ——, and —— (1974*b*). Inelastic behavior of steel frames subjected to constant vertical and alternating horizontal loads, *Proc. Fifth World Conf. Earthquake Eng.*, Rome, **1**, 1194–1197.

——, ——, and I. Mitani (1977). Cyclic behavior of restrained steel brace under axial loading, *Proc. Sixth World Conf. Earthquake Eng.*, New Delhi, **3**, 3181–3187.

—— and K. Minami (1976). Experimental studies on hysteretic characteristics of steel reinforced concrete columns and frames, *Proc. Int. Symp. Earthquake Struct. Eng.*, University of Missouri–Rolla, St. Louis, **1**, 467–480.

—— and —— (1980*a*). Seismic resistance of diagonally reinforced concrete columns, *Proc. Seventh World Conf. Earthquake Eng.*, Istanbul, **6**, 215–222.

—— and —— (1980*b*). Experimental study on the hysteretic characteristics of composite beam-columns, in B. Kato and L.-W. Lu (eds.), *Developments in Composite and Mixed Construction*, Proc. U.S.A.–Japan Seminar on Composite Structures and Mixed Structural Systems, Gihodo Shuppan Co., Tokyo, 197–211.

——, ——, Y. Hisaki, and Y. Miyauchi (1981). Effectiveness of diagonal reinforcement applied to reinforced concrete columns subjected to shear force, *Proc. Third Ann. Conv. Japan Concr. Inst.*, 445–448 (in Japanese).

——, ——, T. Nakamura, R. Sasaki, and S. Morino (1974). Some tests on elastic-plastic behavior of reinforced concrete frames with emphasis on shear failure of columns, *Annuals: Disaster Prevention Res. Inst.*, Kyoto University, no. 17-B, 171–189 (in Japanese).

——, ——, Y. Nishimura, T. Taniguchi, and M. Shimakawa (1978). Experimental study on the elasto-plastic behavior of reinforced concrete frames with multi-story shear walls, *Proc. Arch. Inst. Japan Ann. Conv.*, 1457–1460 (in Japanese).

—— and T. Nakamura (1983). Experimental study on masonry walls, *Proc. Kinki Branch Arch. Inst. Japan*, 461–464 (in Japanese).

——, ——, and H. Matsuda (1977). Experimental study on the stress transmission and load carrying capacity of reinforced concrete beam-to-column connections, *Proc. Arch. Inst. Japan Ann. Conv.*, 1781–1782 (in Japanese).

——, ——, and S. Morino (1973). An experiment of steel reinforced concrete cruciform frames, *Bull. Disaster Prevention Res. Inst.*, Kyoto University, **23**, 75–110.

——, ——, and M. Nakashima (1977). A study on the out of plane inelastic stability of H shaped columns, *Proc. Arch. Inst. Japan Ann. Conv.*, no. 1, 1399–1400 (in Japanese).

——, ——, M. Shibata, and N. Yoshida (1977). Hysteretic behavior of steel braces subjected to horizontal load due to earthquake, *Proc. Sixth World Conf. Earthquake Eng.*, New Delhi, **3**, 3188–3194.

——, ——, and H. Yamamoto (1970). Studies on lateral buckling of wide flange beams, rep. 1, *Annuals: Disaster Prevention Res. Inst.*, Kyoto University, no. 13A, 365–380 (in Japanese).

——, ——, and N. Yoshida (1977, 1980). Experimental studies on the elastic-plastic behavior of braced frames under repeated horizontal loading, parts 1–3, *Bull. Disaster Prev. Res. Inst.*, Kyoto University, **27**, part 3, no. 251, 121–154; **29**, part 3, no. 264, 99–127; **29**, part 4, no. 266, 143–164.

——, ——, ——, and S. Iwai (1978). Effect of strain rate on stress-strain relationships of concrete and steel, *Proc. Fifth Japan Earthquake Eng. Symp.*, Tokyo, 1313–1320.

——, T. Nonaka, and C. Matsui (1969). An experimental study on the horizontal restoring forces in steel frames under large vertical loads, *Proc. Fourth World Conf. Earthquake Eng.*, Santiago, **1**(B-2), 177–193.

———, ———, and S. Morino (1969). An experimental study on the inelastic behavior of steel frames, with a rectangular cross-section subjected to vertical and horizontal loading, *Bull. Disaster Prevention Res. Inst.*, Kyoto University, **18**(145), 65–82.

Watabe, M., and K. Kawashima (1971). A study on shear strength of wooden walls, *Rep. Build. Res. Inst.*, Tokyo, no. 59.

Williams, D., and J. C. Scrivener (1974). Response of reinforced masonry shear walls to static and dynamic cyclic loading, *Proc. Fifth World Conf. Earthquake Eng.*, Rome, **2**, 1491–1494.

Yarimci, E., B. P. Parikh, L.-W. Lu, and G. C. Driscoll, Jr. (1965). *Proposal for Unbraced Multi-Story Frame Tests*, Fritz Eng. Lab. Rep. 273.25, Lehigh University, Bethlehem, Pa.

4

EARTHQUAKE-RESISTANT DESIGN OF BUILDING STRUCTURES

4.1 Design Approaches

4.1.1 Methods of Analysis

Methods of analysis used for the earthquake-resistant design of building structures are classified into static analysis and dynamic analysis.

4.1.1.1 Equivalent-Lateral-Force Procedure
As will be discussed in Sec. 4.2, the equivalent-lateral-force procedure is a method that replaces seismic lateral force by equivalent static lateral force for simplicity in computation. It had long been common practice to assume lateral forces of a constant K times the weight of each element of the structure. More recently there has been a move to use the concept of seismic-base shear, whereby the structure is designed to resist a force applied at the ground which is equal to a constant C_s times the total weight of the structure and which is transmitted to each story of the structure. In Sec. 4.2.2 it will be seen that C_s varies between 0.05 and 0.2 and depends on regional and geological conditions, importance, natural period, ductility and stiffness distribution of the structure, and other factors.

4.1.1.2 Dynamic-Analysis Procedure   Since the lateral force acting on a structure during an earthquake cannot be evaluated precisely by the equivalent-lateral-force procedure, dynamic analysis is adopted when a more accurate evaluation of seismic force and structural behavior is required. Dynamic analysis allows the response of a statically designed structure under dynamic force to be determined and a judgment to be made on the safety of the structure's response. If the response is unsafe, the design is modified to satisfy the required performance of the structure (see Fig. 4-1; Muto, 1974). In this case, the first step of the static design plays an important role.

There are both elastic and inelastic methods of dynamic analysis, but the former are more often used for reasons of simplicity.

Elastic Dynamic Analysis.   The elastic response of a structure under seismic force can best be determined by modal analysis, as explained in Sec. 2.2.4. Time histories of the response of each characteristic mode are first obtained and then summed to obtain the time-history response of the

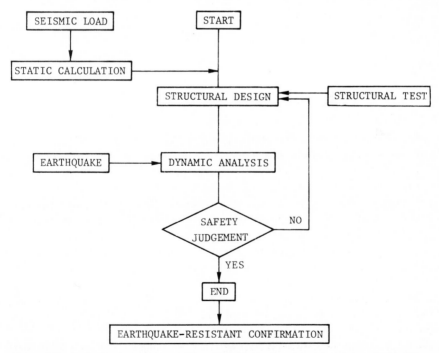

Figure 4-1   Aseismic-design process. [*From K. Muto, Earthquake resistant design of tall buildings in Japan, in J. Solnes (ed.),* Engineering Seismology and Earthquake Engineering, Noordhoff International Publishing, Leiden, 203–245 (1974).]

lumped-mass, n-degrees-of-freedom system. This procedure is called *time-history modal analysis.* Time-history analysis is not always needed, as often only maximum values of response are required for seismic design. In such cases, maximum values of response for each mode are obtained from design spectra and added to determine the maximum response of the whole system. This procedure is called *response-spectrum modal analysis.* The technique of summation generally used is the root-sum-square method. However, the root-sum-square method cannot be used when, for example, there are modes of translational or torsional vibration which have periods nearly equal to the natural period so that coupling occurs. In such cases, direct integration of the equation of motion is required.

Inelastic Dynamic Analysis. To obtain the dynamic response of a structure subjected to a large earthquake, an inelastic dynamic analysis is necessary. Modal analysis can be extended to treat the inelastic range of the response (Chopra and Newmark, 1980). However, for a more rigorous solution, direct integration of the equation of motion is required by using the inelastic restoring-force characteristics given in Sec. 2.5.1 and applying the step-by-step technique mentioned in Sec. 2.6.2. When carrying out this procedure, an appropriate seismic wave must be used as input data. Seismic waves for analytical input data will be discussed in Sec. 4.3.

4.1.2 Selection of Analysis

In the analysis of behavior of structures under seismic force, the more rigorous the analysis, the more reliable and economical the design. It is rational, however, from the engineering viewpoint, to carry out an analysis appropriate to the structural system, configuration, size, importance, and other relevant characteristics of the structure under consideration.

Small single-story wooden and masonry houses can be safely designed simply by specifying, on the safe side, lateral-load-carrying elements per unit floor area, their arrangement, structural detailing, and related matters. This concept is adopted in code rules.

For seismic design of medium-size structures, the equivalent-lateral-force procedure defined by the code is generally used. It is also advisable to check the design with design spectra which correspond to the situation of the structure.

Modal analysis is used for comparatively large and important structures. It should also be used for structures with nonuniform vertical distribution of stiffness or mass, so that modes are superposed to obtain the appropriate vibrational response. In contrast, the static technique uses the first mode of vibrational response of each story, assuming that the vertical distribution of stiffness and mass is the usual distribution.

For very large buildings or very important and potentially dangerous structures, inelastic dynamic analysis is often used to ensure safety when these structures are subjected to severe earthquakes.

For both elastic and inelastic dynamic analysis, a simple lumped-mass system is usually considered as a dynamic model for analysis. However, more complex models must be considered for the analysis of structural frames subject to torsional vibration or other complicated vibration owing to a lack of diaphragm action of the floor.

4.2   Equivalent-Lateral-Force Procedure

4.2.1   Seismic-Base Shear

Almost all building codes adopt the static method of analysis because of the simplicity of its application to seismic design. The seismic-base shear V given by Eq. (4-1) is thus assumed to act on the base of the structure:

$$V = C_s W \tag{4-1}$$

where C_s = the seismic design coefficient and W = the total gravity load of the building. The seismic design coefficient C_s is determined according to locality, natural period, ductility, and other factors as will be described in Sec. 4.2.2. The total gravity load W includes the weight of both non-structural and structural members and also the live load, which is some-what reduced from the value used for vertical-load design.

The base shear V is distributed to each story by the method described in Sec. 4.2.3 to compute the seismic lateral force acting on each story. The stresses induced by this lateral force and the vertical loads are superimposed to check for structural safety (see Sec. 4.6.2.1). Either the allowable-stress design or the ultimate-strength design technique can be used for the safety check (see Sec. 4.6.1).

4.2.2   Seismic-Design Coefficient

Many design codes, such as the Uniform Building Code (UBC, 1979), define the seismic-design coefficient as the product of several other factors (IAEE, 1980). For example, UBC uses the following:

$$C_s = ZIKCS \tag{4-2}$$

where Z = coefficient dependent upon the zone
$\quad\quad\quad I$ = occupancy-importance coefficient
$\quad\quad\quad K$ = coefficient that reflects the type of construction, damping, ductility, and/or energy-dissipation capacity

C = seismic-response factor
S = coefficient for site-structure resonance

The design seismic force should vary with the intensity of earthquakes expected in the area under consideration and is often given by a zoning map. In the case of UBC, the zoning map is as shown in Fig. 4-10 below, and the zoning coefficient is equal to three-sixteenths for Zone 1, three-eighths for Zone 2, three-fourths for Zone 3, and 1 for Zone 4. The UBC gives the occupancy-importance coefficient as 1 for ordinary buildings, 1.25 for buildings which contain over 300 people in one room, and 1.5 for essential facilities. The coefficient K, dependent on damping and ductility, is given by the UBC as 0.67 for ductile-moment space frames, 0.8 for buildings with a dual bracing system, 1.3 for a box system, and 1.0 for other framing systems. The seismic-response factor C depends on design spectra so that the design seismic force varies with the natural period of the building (see Fig. 4-2). The coefficient S takes care of resonance between the building and the ground profile, and the UBC defines it as the ratio of the natural period of the building to that of the ground.

Seismic design coefficients C_s from various countries are summarized in Table 4-1. Almost all countries adopt a similar definition for the coefficient. Also given in the table is the highest value of C_s for each country, which is for a reinforced-concrete building frame constructed in the highest-intensity zone. There are great differences in intensity from country to country. Direct comparison of seismic forces is not possible since some countries use allowable-stress design while others use ultimate-strength design.

4.2.3 Vertical Distribution of Seismic Forces and Horizontal Shear

The variation in seismic force with height in a building is obtained by the superposition of various vibrational modes. It varies with the earthquake-response spectrum, natural period of the building, and vertical distribution of masses and stiffnesses in the building. Ordinary, low- to mid-rise

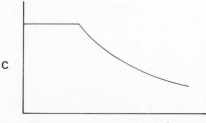

C

Natural undamped period of structure T

Figure 4-2 Relationship between the seismic-response factor and the natural period of a building.

buildings vibrate in the first mode, and seismic forces are assumed to be distributed in the same shape.

According to ATC-3 [1978; see Eq. (2-108)], the seismic design forces F_x applied to each story are as follows:

$$F_x = C_{vx}V \tag{4-3}$$

$$C_{vx} = \frac{w_x h_x^k}{\sum_{i=1}^{n} w_i h_i^k} \tag{4-4}$$

in which W = total gravity load at the building

w_i, w_x = portion of W located at or assigned to level i or x

h_i, h_x = height above the base to level i or x

n = level which is uppermost in the main portion of the building

$k = 1$ for $T \leq 0.5$ s

$k = 2$ for $T \geq 2.5$ s

$k = 0.75 + 0.5T$ for $0.5 < T < 2.5$

Buildings with a short natural period vibrate in a triangular shape similar to the one illustated in A of Fig. 4-3a, while those with a relatively long period vibrate in a more parabolic shape as shown in B of Fig. 4-3a owing to the existence of higher-order vibrations. The distribution of forces is therefore as shown in Fig. 4-3b. Equation (4-3) reflects this condition. The UBC adopts the following equations for horizontal forces, which consist of a concentrated load applied at the top of the building and a triangular-shaped distributed load as shown in C of Fig. 4-3b:

$$V = F_t + \sum_{i=1}^{n} F_i \tag{4-5}$$

$$F_t = 0.07TV \leq 0.25V \tag{4-6}$$

$$F_x = \frac{(V - F_t)w_x h_x}{\sum_{i=1}^{n} w_i h_i} \tag{4-7}$$

where F_t = the portion of V concentrated at the top of the structure and F_x = the lateral force distributed over the height of the structure.

Canada, New Zealand, and Romania use the same equation. Japan, on the other hand, has adopted the distribution illustrated in D of Fig. 4-3b; A of Fig. 4-3b has been adopted by Italy.

TABLE 4-1 Comparison of Earthquake Forces Specified by Codes of Several Countries

Countries and codes	$C_s = ZIKCS$					Examples of C_s^a	Remarks
	Z	I	K	C	S		
Canada	A $0.02-0.08$	I $1.0-1.3$	K $0.7-3.0$	S $0.5/T^{1/2} \leq 1$	F $1.0-1.5$	0.04	$FS \leq 1.0$
Chile	1	$0.8-1.2$	$0.8-1.2$	$0.10 T T_s/(T^2 + T_s^2) \leq 1.0$			$T_s = 0.20-0.90$
China	α_{max} $0.23-0.90$		C $0.25-0.50$	α/α_{max} $0.2 \leq 0.2/T - 0.7/T \leq 1.0$		0.11	
Germany, West	α_0 $0.025-0.10$		α $0.5-1.0$	β $0.528/T^{0.8} \leq 1.0$	K $1.0-1.4$	0.07	
India	α_0 $0.01-0.08$	I $1.0-1.5$		C $0.2-1.0$	β $1.0-1.5$	0.06	
Italy	$0.01(S-2)$		β $1.0, 1.2$	R $0.862/T^{2/3} \leq 1$	ε $1.0, 1.3$	0.10	$S \geq 2^b$
Japan	$C_0 Z$ 0.2 $0.7-1.0$ $^d 1.0$	I 1	D_s 1 $0.25-0.55$	R_T $1 - 0.2(T/T_s - 1)^2$ $1.6 T_s/T$	for $T < T_s$ for $T_s \leq T < 2T_s$ for $2T_s \leq T$	0.20 0.30	$T_s = 0.4-0.8$

Country		C	I	SM		f	
New Zealand		[c]	1.0 – 1.6	0.8 – 2.5 0.8 – 1.2	[c]	0.12	$CS \le 0.14$
Romania		K_c 0.07 – 0.32	ψ 0.15 – 0.35	β 0.75 ≤ 3/T ≤ 2.0	[c]	0.08	
United States	UBC	Z 3/16 – 1.0	I 1.0 – 1.5	K 0.67 – 2.5	C $1/15\,T^{1/2} \le 0.12$ S (≥ 1.0) $1 + T/T_s - 0.5(T/T_s)^2$ for $T/T_s \le 1.0$ $1.2 + 0.6T/T_s - 0.3(T/T_s)^2$ for $T/T_s > 1.0$	0.09	
	ATC-3	A_v 0.05 – 0.40	$1/R$ 0.125 – 0.8	$1.2/T^{2/3}$	S 1.0 – 1.5	0.11	
U.S.S.R		K_s 0.025 – 0.1	K_0 0.5 – 2.0[g]	$\beta = 1/T_i$ 0.8 – 3.0	[g] 0.5 – 2.0	0.10	
Yugoslavia		K_s 0.025 – 0.10	K_0 0.75 – 2.0	K_p 1.0 – 2.0	K_d 0.5/T – 0.8/T	0.10	

NOTE: T = fundamental period of the building; T_s = characteristic site period.

[a] In the example, it is assumed that Z is maximum, I = 1, reinforced-concrete moment-resisting frame, hard ground, and T = 0.5 s (five-story).
[b] S is the seismic degree.
[c] For a moderate earthquake.
[d] For a severe earthquake; ultimate horizontal load-carrying capacity is investigated.
[e] A, C, and S are shown by a figure in a mixed shape.
[f] In addition to the factors shown, risk factor R is multiplied.
[g] Instead of multiplying the factors, the design intensity rating is changed.

Figure 4-3 Vibration modal shapes and distribution of horizontal force along the height. *(a)* Vibration modal shape. *(b)* Horizontal force distribution.

V_x, the story shear at level x, can be computed as the sum of the seismic force acting on all stories above level x (ATC-3, 1978):

$$V_x = \sum_{i=x}^{n} F_i \tag{4-8}$$

4.2.4 Overturning Moment

Overturning moment is created in each level of a building by horizontal force. This produces additional longitudinal stresses in the columns and walls and additional upward or downward forces in the foundations.

The overturning moment at one story of a building can be calculated as the product of horizontal force and the distance from where the force acts to the story under consideration, summed for all the stories above that story. However, the magnitude of the overturning moment can be slightly reduced for the following reasons. First, shear force due to an earthquake as described in Sec. 4.2.3 actually is an envelope of maximum shear forces induced during vibration, and in reality maximum shear forces will never occur simultaneously. Second, severe distress does not occur because of uplift because moment is reduced with uplift as a result of the decrease in stiffness and the consequent decrease in induced force.

ATC-3 introduces the following reduction coefficient (ATC-3, 1978):

$$M_x = \kappa \sum_{i=x}^{n} F_i(h_i - h_x) \tag{4-9}$$

where $\kappa = 1.0$ for the top 10 stories
$\kappa = 0.8$ for the twentieth story from the top and below, and the κ value is interpolated for the stories between the tenth and the twentieth from the top
$\kappa = 0.75$ for the uplift of foundations

4.2.5   Twisting Moment

When the center of stiffness and the center of mass do not coincide at each story, twisting moment is induced in the building and torsional vibration occurs. Even when geometric eccentricity is not present, accidental torsion can occur owing to the torsional component of ground motion, errors in stiffness calculation, eccentricity in live loads, and other unpredictable reasons. In many design codes, eccentricity is therefore assumed to be equal to plus or minus 5 percent of the building width perpendicular to the direction of seismic-force action, in addition to the calculated eccentricity.

Twisting moment is thus obtained statically. However, it is magnified by the action of dynamic force, as described in Sec. 2.4.5. The magnification factor is affected by the torsional rigidity of the building but has not yet been clarified. Design codes today do not mention the magnification factor.

4.2.6   Vertical Seismic Load and Orthogonal Effects

The magnitude of the vertical component of ground motion is often about one-third of the horizontal component. However, in normal design only the horizontal component is taken into account in checking building safety against seismic force. This practice is justified by the fact that a member designed for the horizontal component alone also has adequate resistance to vertical motion. Furthermore, the safety factor used in the design for vertical loading provides a margin which can be taken up and ensures added resistance against the vertical seismic forces of ground motion. However, the vertical component of ground motion may sometimes be as much as two-thirds of the horizontal component. The amplification phenomenon can then threaten the safety of structural members, and the vertical component of ground motion must be taken into consideration in structural design. In this case, design spectra for the horizontal component of ground motion can also be used. Generally, cantilever beams and prestressed-concrete beams are the members that are most strongly affected by the vertical component. For the design of these members, it is recommended that an additional upward force equal to 20 percent of the self-weight of the members be applied (ATC-3, 1978).

The two horizontal components of ground motion are usually applied separately in the design of a building. However, sometimes superposition of the two must be considered. In such cases, resultant force equal to 100 percent of one horizontal component plus 30 percent of the other component is applied to the building. When we consider the effect of earthquake components in two horizontal directions and the vertical direction, we can assume 100 percent of the effect of the input motion in one direction combined with 30 percent of those in the other two directions. More generally, we may use Eq. (4-10) (Chopra and Newmark, 1980):

$$F = (F_1^2 + F_2^2 + F_3^2)^{\frac{1}{2}} \tag{4-10}$$

where F_1 = effect of the component of ground motion in one direction
F_2 = effect of the component of ground motion in a transverse direction
F_3 = effect of the vertical component of ground motion

4.2.7 Lateral Deflection

Since excessive story drift causes failure or inconvenience in secondary (nonstructural) members such as partitions, shaft and stair enclosures, and glass panes, story drift is usually limited to a specified value. For instance, ATC-3 specifies story-drift limit as one-hundred-fiftieth of story height for important buildings and one-hundredth for others, while the UBC and the Japanese code specify a limit of one two-hundredth of story height unless the design demonstrates that a greater drift is acceptable.

When two buildings are located very close to each other or when seismic expansion joints exist, it is necessary to leave enough space between neighboring structures so that they do not pound each other. Extra distance is usually provided between two adjacent buildings in addition to the summation of calculated lateral displacement of both buildings at the nearest possible level of the buildings. In computing lateral displacement, it is necessary to consider plastic deflection, soil-structure interaction, $P\text{-}\Delta$ effect, and other factors besides elastic deflection.

In the computation of lateral deflection, secondary members are considered to be unreliable, and their contribution is usually neglected. Expansion and contraction of columns in slender buildings when bent have a relatively large effect on story drift as well as on total lateral deflection and must be considered.

When checking lateral deflection in the elastic state, elastically calculated deflection can be used. However, since the ultimate state often enters the picture when maximum possible lateral deflection is considered, many design codes recommend that plastic deflection be taken into account. This deflection is equal to elastic deflection multiplied by an amplification factor. The amplification factor is often defined as the inverse of the ductility coefficient K, given with values equal to or close to those in Table 4-1.

4.2.8 P-Δ Effect

When a building in a laterally deflected configuration is subjected to a vertical load, the so-called $P\text{-}\Delta$ effect due to the vertical load P and the lateral deflection Δ induces additional lateral deflection over and above the lateral deflection due to lateral load alone. There are also additional stresses due to the P-Δ effect (see Sec. 3.4.7.1).

The effect of the P-Δ moment at each story is usually evaluated by means of the stability coefficient θ, defined as the ratio of the P-Δ moment to the story moment due to lateral loading:

$$\theta = \frac{P\Delta}{VhC_d} \tag{4-11}$$

where P is the total gravity load applied on the stories above the story in consideration, C_d is the amplification factor described in Sec. 4.2.7, and Δ, V, and h are the design-story drift, shear, and height of the story, respectively. ATC-3 says that the P-Δ effect can be neglected in design if $\theta \leqslant 0.1$ (ATC-3, 1978). If $\theta > 0.1$, the design code requires either that precise calculations be carried out (Tall Building Committee 15, 1979; Tall Building Committee 16, 1979) or that overturning moment and story drift be determined by using a story shear equal to $1/(1 - \theta)$ times V instead of V (ATC-3, 1978).

4.2.9 Soil-Structure Interaction

In determining the design seismic force, it is usual to assume that ground motion at the foundation of a building is equal to free-field ground motion, i.e., to the motion which would occur if no building were in existence on the site. Precisely speaking, this assumption holds only when the ground is rigid. When the ground is soft, the natural period of the building is apt to become longer since the foundation motion contains both rocking and translational components. Furthermore, the greater part of the seismic energy is consumed by radiation (or geometric) damping during radiation of seismic waves away from the foundation and also by damping of the soil material as a result of inelastic hysteretic action in the soil. The seismic forces, i.e., base shear, lateral force, overturning moments, etc., consequently tend to become smaller. Lateral displacement and the P-Δ effect, on the other hand, tend to become larger.

A simplified method of calculating the modified values of base shear and lateral deflection in such cases is given in ATC-3.

4.3 Design Earthquakes

4.3.1 Seismic-Hazard Study

Although aseismic design is usually carried out in accordance with relevant design codes, it would be preferable to base such design on predicted seismic hazard at the site. This procedure is in fact followed in the design of large-scale, important structures.

The seismic hazard of a building site is determined from the following factors:

1. Geographical factors such as pattern, type, and movement of a nearby active fault or faults and distance from such a nearby fault or faults

2. Seismicity data such as the distribution of epicenters of past earthquakes

3. Isoseismal maps (Modified Mercalli, or MM), which tell us much about the intensity, the grade of ground motion, and the distribution of future earthquakes

4. Geological features such as soil-profile densities, shear wave velocities, and shear moduli (Cherry, 1974*a*, 1974*b*, 1974*c*).

4.3.2 Earthquake Records for Design

When the hazard study has been completed and the expected maximum seismic acceleration obtained, dynamic analysis proceeds for a suitable seismic wave chosen from the records of strong earthquakes in the past. Among strong-earthquake records, El Centro (1940) and Taft (1952) are famous and are often used in conjunction with other records of earthquakes in the neighborhood of the site in question. These earthquake records are used in dynamic analysis with modifications to fit site conditions. Generally, the amplitude of acceleration is adjusted to that expected at the site. It is also recommended that frequency content and duration be adjusted according to the magnitude and MM intensity (Seed, Idriss, and Kiefer, 1969).

4.3.3 Factors Affecting Accelerogram Characteristics

Earthquake motions are characterized by the following parameters: (1) maximum amplitude of acceleration, (2) predominant frequency or period of motion, and (3) duration of motion. From these characteristics the time history of motion can be obtained. To this end, either a synthetic earthquake motion with desired characteristics is generated or an existing earthquake is modified (Seed, Idriss, and Kiefer, 1969).

The duration of shaking is also an important earthquake characteristic. As indicated by Eq. (1-10), the rupture of a fault lengthens when earthquake magnitude increases. On the other hand, the duration of shaking is longer for earthquakes of larger magnitude, since the velocity at which rupture propagates along a fault is approximately 3 km/s and constant.

The relation between the magnitude and the duration of earthquake motion, as obtained from earthquake records, is shown in Fig. 4-4 (Housner, 1965).

The frequency content of waves changes with the distance from the causative fault to the site. In other words, high-frequency waves attenuate most significantly in crustal rock, and the predominant period lengthens when the site is more distant from the fault. Figure 4-5 shows the relation between the predominant period and the distance from the causative fault for different values of magnitude, as obtained from measured data (Seed, Idriss, and Kiefer, 1969).

It is obvious from the definition given in Sec. 1.1.3.2 that the maximum amplitudes of acceleration in rock increase with an increase in magnitude and attenuate with an increase in distance from the earth-

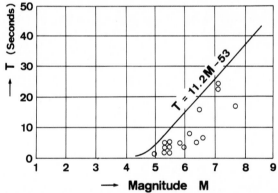

Figure 4-4 Relationship between the magnitude and the duration of earthquakes. [*From G. W. Housner, Intensity of earthquake ground shaking near the causative fault,* Proc. Third World Conf. Earthquake Eng., *Auckland, 1(3), 94−111 (1965).*]

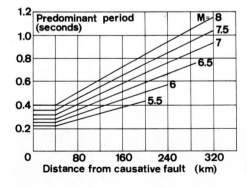

Figure 4-5 Predominant periods for maximum acceleration in rock. [*From H. B. Seed, I. M. Idriss, and F. W. Kiefer, Characteristics of rock motions during earthquakes,* J. Soil Mech. Found. Div., Am. Soc. Civ. Eng., *95(SM-5), 1199−1218 (1969).*]

quake focus. Gutenberg and Richter (1956), Esteva and Rosenblueth (1963), Blume (1977), and others propose some method of determining the maximum amplitude of acceleration in rock. In Seed, Idriss, and Kiefer (1969), the variation of maximum acceleration in rock with earthquake magnitude and distance from the causative fault is obtained, as Fig. 4-6, by averaging the above data. Trifunac and Brady (1975) compare various proposals in addition to those mentioned above. More recently, various empirical formulas have been proposed on the basis of recently observed data (Donovan, 1974; Blume, 1977; McGuire, 1977; RCDERT, 1977; Blume, 1980). Housner (1965) discusses the upper bound for the intensity of ground shaking on reasonably firm and deep alluvium. Housner (1969) also gives idealized relations, for California, between fault length and magnitude, the special distribution of intensity of ground shaking and magnitude, the affected area and magnitude, and frequency of occurrence and magnitude.

The influence of geological formations is scarcely considered in earthquake records in the United States. However, in Mexico City, where a thick, soft layer of volcanic ash exists on top of bedrock, and in many cities in Japan, where thick and soft alluvial formations exist, the amplification effect of earthquake waves propagating from bedrock to the earth surface cannot be neglected, and seismic waves must be modified accordingly. When modification is required, either the method of replacing the soft

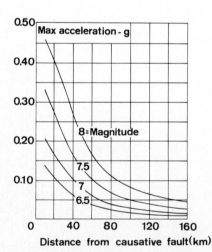

Figure 4-6 Variation of maximum acceleration with earthquake magnitude and distance from the causative fault. [*From H. B. Seed, I. M. Idriss, and F. W. Kiefer, Characteristics of rock motions during earthquakes,* J. Soil Mech. Found. Div., Am. Soc. Civ. Eng., **95**(SM-5), 1199–1218 (1969).]

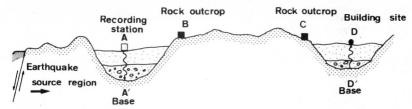

Figure 4-7 Schematic representation of a procedure for computing the effects of local soil conditions on ground motion.

ground by a lumped mass or wave-propagation theory can be used to obtain design seismic waves.

To predict earthquake motion at site D from the known earthquake record at site A (Fig. 4-7), the motion of the base rock A' immediately under site A is first obtained, and then that of the nearest rock outcrop B. This motion is modified by considering the distance from the focus, and the motion at the rock outcrop C near site D is then computed. The earthquake wave at site D is thus determined from the motion of the base rock D' immediately under site D (Schnabel, Seed, and Lysmer, 1971). Several computer programs are available to follow this procedure, and they have been applied in practical designs. However, it has been suggested that the procedure should not be relied on until many earthquake records, taken simultaneously at bedrock, at the rock outcrop, and at ground surface, demonstrate its adequacy, since the procedure involves only very simple assumptions.

Kanai derived empirical formulas from earthquake records taken at both bedrock and the ground surface in Japan and the United States (Kanai, 1960; Kanai, 1966). These formulas express the spectral displacement, spectral velocity, and spectral acceleration of bedrock motion as a function of earthquake magnitude M and hypocentral distance x. They also express the spectra of motion on the ground surface in terms of T_G, the predominant period of the ground, together with M and x. Figure 4-8 illustrates the spectral displacement, velocity, and acceleration of motion on the ground surface by using the above-mentioned formulas.

4.3.4 Artificial Accelerogram

Earthquake records taken on the ground are greatly affected by the source mechanism, travel-path geology, local site conditions, and other factors, and as records are limited in number, artificial (simulated) ground motions are sometimes used for dynamic analysis in addition to the records of actual earthquakes. For instance, when designing nuclear power plants in the United States and Japan for earthquake forces, artificial

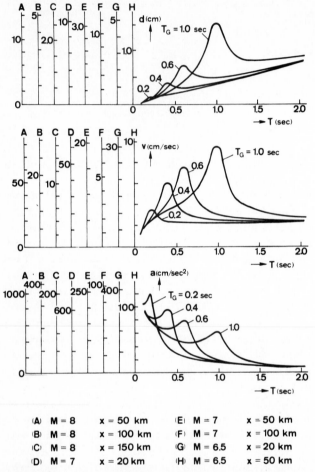

(A)	M = 8	x = 50 km	(E)	M = 7	x = 50 km
(B)	M = 8	x = 100 km	(F)	M = 7	x = 100 km
(C)	M = 8	x = 150 km	(G)	M = 6.5	x = 20 km
(D)	M = 7	x = 20 km	(H)	M = 6.5	x = 50 km

Figure 4-8 Spectra of displacement, velocity, and acceleration of ground surface. [*From K. Kanai, Improved empirical formula for the characteristics of strong earthquake motions,* Proc. Japan Earthquake Eng. Symp., *Tokyo, 1–4 (1966).*]

seismic waves are used as well as the records of actual earthquakes. Artificial seismic waves are constructed statistically from frequency content, amplitude variation, and shaking duration. They are convenient since relatively continuous response spectra which well match the smooth design spectra can be obtained.

Artificial seismic waves proposed so far (Bycroft, 1960; Tajimi, 1960; Housner and Jennings, 1964; Jennings, Housner, and Tsai, 1969) fall roughly into three categories according to the methods used (Toki, 1981):

1. Method of composing many harmonic waves that have different amplitudes and phase angles

2. Method of using the response spectrum as ground motion, which is obtained when a linear system of one degree of freedom is subjected to white noise

3. Method of distributing various pulses randomly along the time axis to compose a new wave

Of these three methods, the first is most often used. In this method, the seismic-acceleration spectrum is expressed as a product of a function that has prescribed power spectral characteristics and an envelope function that represents the change in amplitude with the elapse of time. As an envelope function, the one shown in Fig. 4-9, which is associated with earthquake magnitude and others, has been proposed (Jennings, Housner, and Tsai, 1969). The resulting artificial wave is further modified so that the response spectrum for the simulated wave is close to the smooth response spectrum given by design codes.

4.3.5 Zoning Map

Since the earthquake motion to be expected at a site varies vastly, a zoning map which gives an idea of the size of earthquakes is required for both static and dynamic analysis. One zoning map is a seismic map which provides a record of past earthquakes and predicts possible future earthquakes. Simple zoning maps in Figs. 1-12, 1-14, 1-15, and 1-16 give the size and epicenters of historical earthquakes. While seismicity maps contain helpful information for seismic design, engineering maps can be directly used for design. An engineering map gives the magnitude of the design ground motion itself.

The zoning map for the United States, as given in the Uniform Building Code (IAEE, 1980), is shown in Fig. 4-10. There can be different engineering maps, depending on design objectives, e.g., whether the

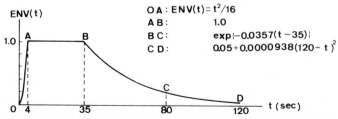

Figure 4-9   Envelope function for earthquake type A. [*From G. W. Housner and P. C. Jennings, Generation of artificial earthquake,* J. Eng. Mech. Div., Am. Soc. Civ. Eng., *90(EM-1), 113–150 (1964).*]

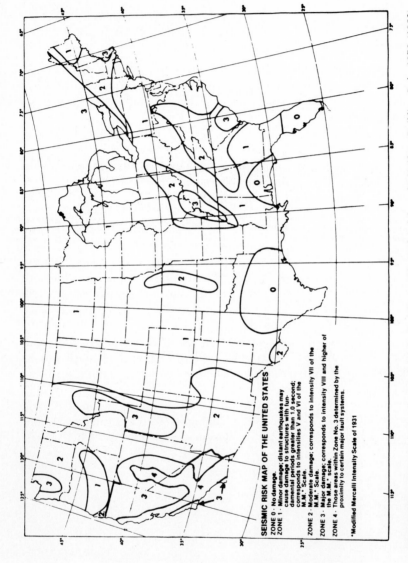

SEISMIC RISK MAP OF THE UNITED STATES

ZONE 0 · No damage.
ZONE 1 · Minor damage; distant earthquakes may cause damage to structures with fundamental periods greater than 1.0 second; corresponds to intensities V and VI of the M.M.* Scale.
ZONE 2 · Moderate damage: corresponds to intensity VII of the M.M.* Scale.
ZONE 3 · Major damage: corresponds to intensity VIII and higher of the M.M.* scale.
ZONE 4 · Those areas within Zone No. 3 determined by the proximity to certain major fault systems.

*Modified Mercalli Intensity Scale of 1931

Figure 4-10 Seismic zoning map of the United States. [*Reproduced from the Uniform Building Code, 1979 (1982) edition, copyright © 1979 (1982), with permission of the publisher, The International Conference of Building Officials (UBC, Whittier, Calif., 1979).*]

importance of the building and the danger to human life are of first priority or whether economic loss due to damage is considered first. Generally, a map for one purpose should not be used directly for another purpose (Housner and Jennings, 1974).

4.4 Dynamic-Analysis Procedure

4.4.1 Modal Analysis

When carrying out a dynamic analysis, it is usual to replace the mass of each story by a lumped mass at each floor level. There exist as many vibrational modes as the number of masses. However, for simplicity in computation the first three modes are usually considered for low- to middle-rise buildings, and six modes for high-rise buildings. For the computation of vibrational modes and natural periods, ready-made computer programs are often utilized, as mentioned in Sec. 2.5.3.

The base shear V_n for the nth mode is obtained from Eqs. (2-45) and (2-105) as follows:

$$V_n = \frac{S_{an}}{g} \bar{W}_n \tag{4-12}$$

$$\bar{W}_n = \frac{\left(\sum\limits_{i=1}^{N} w_i \phi_{in}\right)^2}{\sum\limits_{i=1}^{N} w_i \phi_{in}^2} \tag{4-13}$$

in which S_{an} = ordinate corresponding to the nth natural period of the pseudo-acceleration response spectrum and damping ratio

g = acceleration of gravity

w_i = weight lumped at the ith level

ϕ_{in} = displacement amplitude at the ith level when vibrating in its nth mode

N = total number of floors

The term S_{an} can be obtained from the design spectrum. The value S_{an}/g corresponds to the seismic design coefficient C_s given by Eq. (4-1). It is determined by ATC-3 as the product of zoning, ductility, and other factors in a form similar to that of the equivalent-force procedure (ATC-3, 1978).

From Eq. (2-108), the horizontal force acting on the ith floor level due to the vibration of the nth mode is

$$F_{in} = V_n \frac{w_i \phi_{in}}{\sum\limits_{i=1}^{N} w_i \phi_{in}} \tag{4-14}$$

The modal deflection at each level δ_{in} is given by the following equation, which refers to Eq. (2-110):

$$\delta_{in} = \frac{g}{4\pi^2} \frac{T_n^2 F_{in}}{w_i} \tag{4-15}$$

The modal drift Δ_n can be computed as the difference between δ_{in} for the floor above and the floor below. The design values are obtained by multiplying these values by the amplification factor C_d. The modal story shear and the modal moment can be obtained statically by applying the horizontal force F_{in} to the masses. The total response, that is, the story shears, overturning moments, drift quantities, and deflection at each level, all of which are to be used in design, can be computed by the root-sum-square method [see Eq. (2-103)], using the above-obtained modal values.

4.4.2 Inelastic-Time-History Analysis

In inelastic-time-history analysis, as in modal analysis, the structure is usually replaced by a lumped-mass system. It is necessary to establish a hysteresis model associated with restoring force, as described in Sec. 2.5.1. After the damping ratio and the hysteresis model have been determined and an appropriate design earthquake for the building site chosen, step-by-step direct integration is carried out by electronic computer. The important results of the computation are the maximum values in the time history of quantities such as story shear, story-shear coefficient, over-turning moment, deflection, story drift, story ductility, and member ductility. These are then checked against prescribed dynamic design criteria following the procedure to be described in the next section. If any inappropriate value is obtained, the design is changed and checks and modifications are repeated until satisfactory results are achieved (see Fig. 4-1).

4.4.3 Evaluation of the Results

As illustrated in Fig. 4-1, the results of dynamic analysis must be evaluated with the dynamic design criteria and should be modified if the criteria are not satisfied. The dynamic design criteria require that the story shear not exceed the allowable value and that the ductility factors of floor level, member, and connection not exceed prescribed values in the case of inelastic response.

Examples of the finally obtained responses of a 21-story steel building to El Centro north-south motions are shown in Figs. 4-11 through 4-13. In this case, the criteria are that the response be below the elastic limit for a moderate earthquake of maximum amplitude of 200 gal, that story shear be smaller than elastic-limit strength, and that the ductility factor be smaller than 2 for a severe earthquake of 350 gal. Also, story drift should not exceed one two-hundredth of story height for earthquakes of 200 gal.

If the response of one portion of the building happens to be much larger than that of other portions, this portion also must be reinforced so as to have a smooth response curve.

4.5 Fundamental Aseismic Planning

In the very early stage of building design, the configuration, basic materials, structure, and framing of the building have to be chosen. These choices greatly affect later aseismic design. The architect and the structural engineer should therefore cooperate and thoroughly discuss the matter at this early stage.

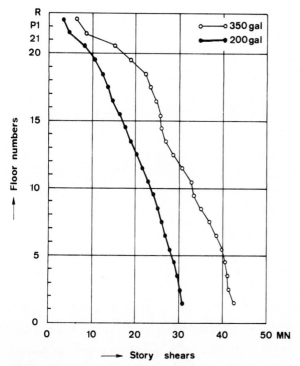

Figure 4-11   Maximum story shears for El Centro north-south motions. *(Courtesy of Nikken Sekkei Co.)*

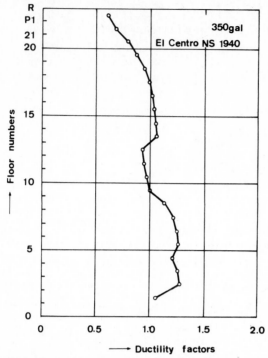

Figure 4-12 Maximum ductility factors for El Centro north-south motions. *(Courtesy of Nikken Sekkei Co.)*

Structural materials have their own performance characteristics and should be selected according to the location and condition of the building to be planned in order to accomplish safe, economical, and superior architecture. There have been many cases in which a building with a configuration chosen by an architect has suffered damage because of inadequate earthquake resistance. Once a building has been given an aseismically disadvantageous configuration, a sound structure cannot be realized even when structural design is properly carried out. This is true also of framing, distribution of load-bearing building elements, and other structural factors. Both architects and structural engineers should therefore be familiar with the effect on earthquake resistance of material characteristics, configuration, and structural framing.

4.5.1 Selection of Materials and Types of Construction

From the standpoint of aseismic design, the following major characteristics should be provided in a building (see also Sec. 2.7):

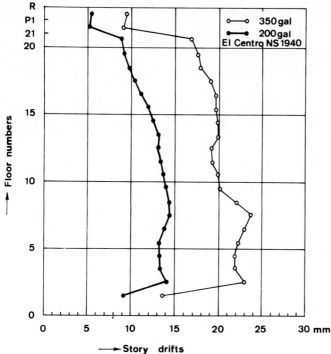

Figure 4-13   Maximum story drifts for El Centro north-south motions. *(Courtesy of Nikken Sekkei Co.)*

1. *High strength-to-weight ratio.* Since seismic force works on a building as an inertial force, it is advantageous to use a light and strong material and/or structural system.

2. *High deformability.* High plastic-deformation capacity can compensate for any shortage in strength.

3. *Low degradation.* It is desirable to use a structural system which displays low degradation in strength and stiffness under repeated loading.

4. *High uniformity.* Separation of structural elements when subjected to earthquakes should be prevented.

5. *Reasonable cost.* A building plan is often discarded because of unreasonably high cost despite its superior physical quality. The economic situation differs from country to country.

Among structural materials such as concrete and steel, the superiority of one material over another can be determined by checking the items listed above. However, since both ductile and brittle members can result from a combination of brittle concrete and ductile steel, performance of structural components cannot be evaluated by materials alone. Further factors, such as structural continuity at connections, must be considered in the evaluation of an entire structural system composed of such structural components for earthquake resistance.

Building structures can generally be given the following priority ordering for earthquake resistance:

a. Steel structures. These are especially suitable for high-rise buildings if the major characteristics of items 1 through 4 above are satisfied. In many countries (excluding the United States and Japan) steel structures often are not adopted for low- to middle-rise buildings because item 5 is not satisfied.

b. Steel and reinforced-concrete composite structures. These have characteristics intermediate between those of pure-steel structures and reinforced-concrete structures. In Japan, they have been used very often for middle- to high-rise buildings.

c. Wooden structures. These are superior in earthquake-resistant characteristics and satisfy all items 1 through 5, although they are inferior in fire resistance. They are often used for low-rise houses.

d. Cast-in-situ reinforced-concrete structures. These are inferior to steel structures with respect to items 1 through 3. To avoid disadvantages related to items 2 and 3, various measures have been proposed and tried. This form of construction is widely adopted in low-, middle-, and high-rise buildings. In Japan, where very large design earthquake forces must be considered, the use of cast-in-situ reinforced-concrete structures is limited, as a rule, to buildings of six or fewer stories.

e. Precast-concrete structures. Concrete structures with precast elements are used for low- to middle-rise buildings. This type of structure lacks uniformity in comparison with the cast-in-situ integral concrete structure. There are not many problems when a building is composed of precast-concrete panels, but when it is composed of linear elements, it is inferior in the aseismic sense because the structural system is not monolithic.

f. Prestressed-concrete structures. The introduction of prestressing into a structural element adversely affects its deformability and hence the aseismic characteristics of the building. The earthquake-resistant characteristics of prestressed-, precast-concrete structures are therefore inferior to those of their prestressed counterparts. Prestressed cast-in-situ concrete structures are also inferior to their unprestressed counterparts. Prestressed-concrete structures are used for middle- and low-rise buildings.

g. Masonry structures. Reinforced masonry is relatively superior with regard to items 1 through 3, that is, strength-to-weight ratio, deformability, and degradation. Earthquake resistance compatible with that of a reinforced-concrete structure can thus be achieved provided the structure is properly designed and constructed. Masonry structures also satisfy item 5 very well. Reinforced-masonry construction is thus adopted for middle- and low-rise buildings in many countries.

Mixed structures which combine two or more of these structural systems can also perform well with regard to earthquake resistance when the proper member is used in the proper position. In mixed structures, it is important to ensure load-carrying capacity and ductility in the interfaces of the various components.

4.5.2 Form of Superstructure

4.5.2.1 Building Configuration

Configuration of Plan. The plan comprises simplicity, compactness, and large torsional rigidity.

Simplicity. From the viewpoint of earthquake resistance, a simple configuration such as a square or a circular shape is desirable. In buildings with winged shapes such as the L, T, U, H, Y, and others shown in Fig. 4-14*a*, the wing portion often collapses under a severe earthquake as illustrated in Fig. 4-14*b*. In such cases, seismic joints which structurally separate the wings, as shown in Fig. 4-15, should be provided. There must be enough clearance at the seismic joints so that the adjoining portions do not pound each other (see Sec. 4.2.7).

Compactness. In a building with a long, extended shape, complicated forces act because of the difference in the phase of the seismic motion (Fig. 4-15*b*). Seismic joints are required in such a building.

Symmetry and large torsional rigidity. To avoid torsional deformation, the center of stiffness of a building should coincide with the center of mass (see Fig. 4-16). To satisfy this condition, it is desirable to have symmetry both in the building configuration and in the structure. Although the center of stiffness can be made to coincide with the center of mass in an asymmetric building, it is often difficult to maintain coincidence in the inelastic state of stress.

If eccentricity exists between the centers of mass and stiffness, torsional deformation and amplification of seismic motion are greater in a building with lower torsional stiffness (see Fig. 4-17 and Sec. 2.4.5).

Vertical Configuration. The vertical configuration comprises uniformity and continuity as well as proportion.

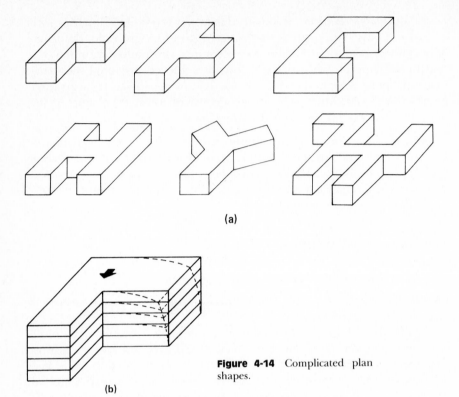

(a)

(b)

Figure 4-14 Complicated plan shapes.

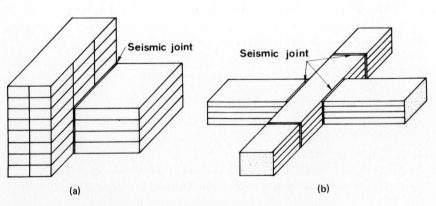

(a)

(b)

Figure 4-15 Seismic joints.

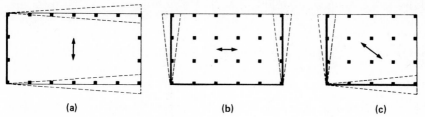

| (a) | (b) | (c) |

Figure 4-16   Structurally asymmetrical plans.

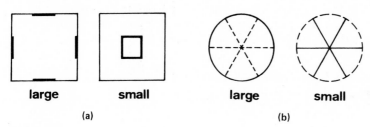

| large | small | large | small |

| (a) | | (b) | |

Figure 4-17   Torsional rigidity.

Uniformity and continuity. It is desirable to avoid drastic changes in the vertical configuration of a building. When the vertical configuration is discontinuous as shown in Fig. 4-18, a large vibrational motion takes place in some portion and a large diaphragm action is required at the border to transmit forces from the tower to the base. In such cases, dynamic-response analysis is mandatory to ensure earthquake resistance.

Proportion. A building with a large height-to-width ratio exhibits large lateral displacement under lateral force. The axial-column force

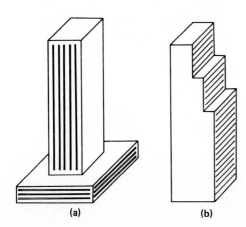

| (a) | (b) |

Figure 4-18   Setback.

due to overturning moment in such a building tends to become un-manageably large. So too do the compressive and the pullout forces which act on the foundation. In Japan, the static earthquake load is increased in the design of buildings with height-to-width ratios larger than 4.

4.5.2.2 Stiffness and Strength

Vertical Direction. It is advisable to avoid sudden changes in the vertical distribution of stiffness and strength. The relevant parameter is the ratio of story stiffness to story weight between adjacent stories. If there is a soft story in a building as illustrated in Fig. 4-19, plastic deformation tends to concentrate in the soft story, and this may cause the entire building to collapse. Such a condition is often found in buildings with pilotis in the first floor and shear walls above. Stiffness and strength can be adjusted by increasing columns or bracings in the soft stories.

Horizontal Direction. If both long and short columns exist in the same story, shear force is concentrated in the relatively stiff short columns, which thus fail before the long columns. In a structural frame, long columns can be turned into short columns by the introduction of spandrels. In this case, nonstructural walls should be separated from structural members (see Sec. 4.7.5). Forces concentrate also in short beams. This situation can be alleviated by adjusting beam depths, hence avoiding stress concentrations. Reinforced-concrete members likely to suffer from stress concentration can be endowed with ductility by diagonal reinforcement (refer to Sec. 3.3.4.2).

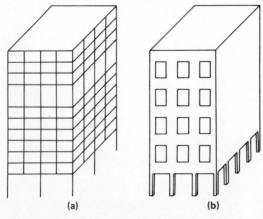

(a)   (b)

Figure 4-19 Soft stories.

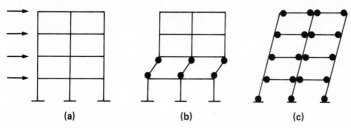

Figure 4-20 Selection of a failure mode. *(a)* Frame. *(b)* Weak columns, strong beams. *(c)* Weak beams, strong columns.

4.5.2.3 Some Other Considerations

Redundancy. Redundancy in structural systems should be kept low in view of thermal stress and uneven settlement of the ground. Under the action of earthquake forces, on the other hand, high redundancy is desirable because local failure does not then induce collapse of an entire building if the capacity for plastic deformation is large.

Failure Mode. Although a structural designer can choose whether the columns or the beams should yield first, it is generally desirable to provide strong columns and to allow prior yielding of the beams in flexure. The reasons for this choice are the following:

1. Column failure means the collapse of the entire building.

2. In a weak-column structure, plastic deformation is concentrated in a certain story, as shown in Fig. 4-20, and a relatively large ductility factor is consequently required.

3. In both shear failure and flexural failure of columns, degradation is greater than in beam yield. This is true because of the axial forces in the columns.

Even when a frame is designed with strong columns and weak beams, plastic hinges form at the base of the lowest-story columns in a statical failure mode as shown in Fig. 4-20c. Plastic hinges also form in many column ends during inelastic vibration. Columns should therefore always be provided with adequate ductility.

Stiff or Flexible Structures. A flexible structure such as a moment-resisting steel frame is advantageous for a site where expected ground motion is of short period, since it takes up a relatively small motion. However, a flexible structure tends to exhibit large lateral deflections which induce damage in nonstructural members. In tall buildings, oscillations due to wind gusts can cause discomfort in the occupants, and a stiff structure is desirable. Therefore, it cannot definitively be said that one type is superior to the other.

4.5.3 Framing Systems and Aseismic Units

4.5.3.1 Framing Systems

Moment-Resisting Frame.   The moment-resisting frame is the fundamental structural system. In reinforced-concrete structures, the moment-resisting frame includes the cast-in-place frame and the precast frame. The former is basically the beam-and-column frame illustrated in Fig. 4-21*a;* it also includes the flat-slab-and-column frame, which is composed of columns and floor slabs, and the slab-and-load-bearing-wall frame. For the flat-slab type of structural system, various measures have been proposed and put into practice to provide sufficient strength and ductility in flat-slab-to-column connections.

For precast reinforced-concrete frames, beam-to-column connections at the site are often welded. However, it is comparatively difficult to achieve uniformity and continuity of the structure. On the other hand, it is much easier to maintain structural uniformity and continuity in cast-in-place connections by site-jointing the members in midspan regions where working stresses are relatively small.

Almost perfect moment-resisting joints have been achieved in steel and steel-concrete composite moment-resisting frames by the use of special fasteners and/or welding techniques. To assure ductile failure under repeated loading, it is advisable to design joints to be stronger than adjoining members.

Frames with Vertical Diaphragms.   If the strength and stiffness of a frame are not adequate, load-bearing walls and/or bracings are often used to complement the frame's strength and stiffness, as illustrated by Fig. 4-21*b* through *e*. Shear walls and bracings are useful also for protecting non-structural components from failure by reducing story drift. Figure 4-21*b* shows the case in which bearing walls are adopted, and *c* shows the combined use of columns and bearing walls. Figure 4-21*d* illustrates the use of infilled walls. From an aseismic point of view, *c* and *d* are superior to *b*. However, *b* is advantageous for flexibility of the enclosed space and thus is often used. Type *d* has often been used in Japan, but it is not common in the rest of the world since use of the interior space is limited. Load-bearing walls are not necessarily used only for reinforced-concrete frames but frequently also for steel or steel-concrete composite frames. Steel bracings as shown in Fig. 4-21*e* are mainly used in steel frames. Reinforced-concrete bracing is seldom used because it is difficult to anchor it to the frame.

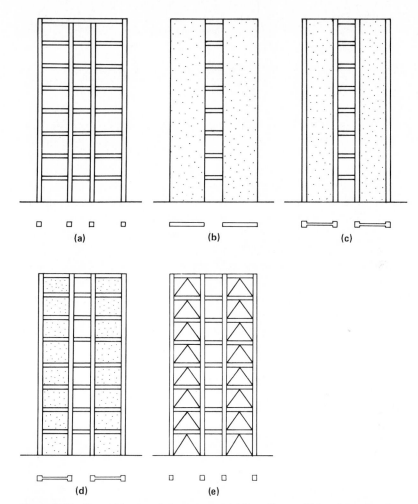

Figure 4-21 Lateral-load-resisting systems. *(a)* Pure frame. *(b)* Load-bearing shear walls. *(c)* Shear walls with columns. *(d)* Infilled shear walls. *(e)* Braced frame.

Vertical diaphragms and bracings are usually placed in internal walls, external walls, or core walls. Core walls are walls encasing elevator shafts, staircases, and other such spaces. Cores are usually located as shown in Fig. 4-22. Structural cores should be located so as to minimize eccentricity between the center of mass and the center of stiffness.

4.5.3.2 Behavior of Frames under Horizontal Forces In a high-rise building frame with load-bearing walls or bracings, the two different structural elements (that is, the frame plus the walls or the bracing) work together to

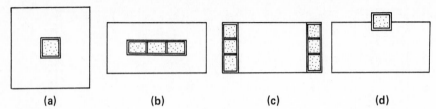

Figure 4-22 Location of cores. *(a)* Center. *(b)* Middle portion. *(c)* Ends. *(d)* Eccentric location.

resist horizontal forces. Shear deformation is predominant in the pure frame (Fig. 4-23*b*), while flexural deformation is predominant in the shear-wall system (Fig. 4-23*c*). A structural frame with load-bearing walls hence exhibits an intermediate form of behavior. In the lower part of the building, the walls resist the greater part of the shear force, but the share gradually decreases in higher stories (Fig. 4-23*d*). Indeed, in the upper portion of the building, the share may become negative. This is the case when rotation of the wall base is prevented. If rotation of the wall anchors is allowed, this trend of the negative shear is more likely to occur.

If flexural deformation occurs in a load-bearing wall, the adjacent boundary beam undergoes large deformation and should have adequate ductility. Also, adjacent columns are subject to large axial force, so that difficulties arise both in designing the column cross section and in dealing with the pullout force on the foundation. To avoid this disadvantageous situation, it is advisable to distribute the structural elements which carry horizontal forces as in Fig. 4-24*a* and *b*. It is also advisable to use a large, rigid frame, as indicated in Fig. 4-24*c* and *d,* by making the walls or bracings deep in some regions. If it is still difficult to handle large axial forces in the columns and large pullout forces on the foundations, the number of frame planes with horizontal load-bearing elements should be increased.

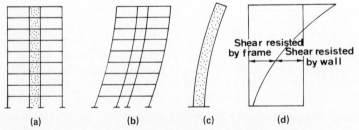

Figure 4-23 Contribution of frames and shear walls to story shear. *(a)* Mixed system. *(b)* Frame. *(c)* Shear wall. *(d)* Share of story shear.

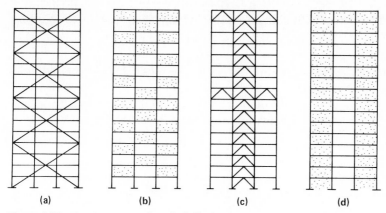

Figure 4-24   Arrangement of vertical diaphragms.

4.5.3.3 Aseismic Units

Reinforced-Concrete Shear Wall. There are two types of reinforced-concrete shear wall: the cast-in-place shear wall and the precast shear wall. In the cast-in-place shear wall, the wall reinforcement is anchored into the surrounding frame to ensure uniformity and continuity. In the precast shear wall, on the other hand, uniformity and continuity are achieved by providing trapezoidal-shaped keys along the panel edges or by connecting panels to the frame with steel joints. Various measures have been proposed and put into practice to provide load-bearing walls with ductility and energy-dissipation capacity. Examples of such measures are given in Fig. 4-25, where *a* shows a wall with slits and *b* indicates a ductile and flexible joint connecting the precast wall to the surrounding frame (Tall Building Committee 21D, 1979).

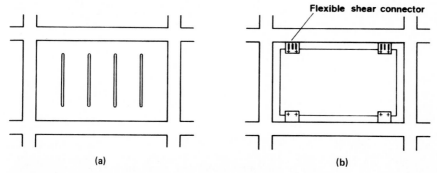

Figure 4-25   Ductile shear walls. *(a)* Slit shear wall. *(b)* Shear wall with flexible shear connectors.

Steel Shear Walls. Sometimes, steel plates are used as shear walls. Vertical and horizontal stiffeners must be provided on such steel shear walls to prevent local buckling.

Composite Shear Walls. Composite shear walls include stiffened steel plates embedded in reinforced concrete, steel-plate trusses embedded in a reinforced-concrete wall, and other possible composite walls, all of which are combined with either a steel frame or a composite frame.

Steel Bracings. When such simple and slender bracings as reinforcing bars or steel plates are used as bracings, it is assumed that they resist no axial compression. Structural shapes such as angles, channels, wide flanges, and tubes are also often used as bracings. In a high-rise building, relatively stocky bracings are generally used to provide the structure with improved hysteretic characteristics.

Masonry Shear Walls. Solid masonry walls without any reinforcement have long been used in masonry buildings. However, they have turned out to be inadequate from the aseismic point of view, and reinforced-masonry walls such as hollow-unit masonry and grouted masonry are now used instead.

4.5.4 Devices for Reducing Earthquake Load

Many devices have been proposed either to prevent an earthquake force from acting on a structure or to absorb a portion of the earthquake energy that is introduced into the structure.

4.5.4.1 Base-Isolated Structures

The schematic drawing in Fig. 4-26 indicates one traditional concept of structural base isolation whereby rollers prevent ground motion from being transmitted from the building foundation into the superstructure. Recently, more practical devices have been proposed and actually used. The following are examples of base-isolated structures:

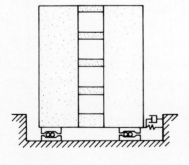

Figure 4-26 Schematic shape of a base-isolated structure.

1. A three-story school building with shear walls was isolated by setting the entire structure on a rubber cushion (Petrovski, Jurukovski, and Simovski, 1978).

2. A building resting on ball bearings was provided with control rods to prevent lateral movement of the building as a result of small lateral forces (Caspe, 1970).

3. In a steel-frame building, stiff columns made of reinforced concrete were placed to adjoin steel columns at the basement. The building can behave as a steel-frame structure under service-load conditions. If a large earthquake force is applied to the building, the reinforced-concrete columns function as stoppers to prevent excessive lateral deflection (Matsushita and Izumi, 1965).

The most practical device so far proposed for base isolation has been developed in New Zealand, where it has been extensively studied and is now covered in code provisions (Blakeley, Charleson, et al., 1979; Lee and Medland, 1978a, 1978b; Tyler, 1977; Megget, 1978; Priestley, Crosbe, and Carr, 1977; Robinson and Tucker, 1977). A horizontal flexible mounting supports the building while allowing it to move freely in the horizontal direction. A load limiter resists movement due to small horizontal loads (see Fig. 4-27). The horizontal flexible mounting is either a sandwiched rubber-and-steel plate or PTFE (Teflon). Up to a certain yield load the load limiter works elastically, and at higher loads it exhibits plastic deformation, which allows horizontal building movement. The

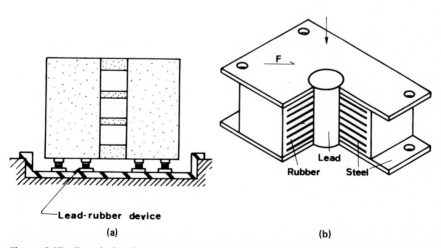

Figure 4-27   Base-isolated structures in New Zealand. (a) Idealized elevation of a base-isolated building. (b) Lead-rubber device. [*From R. W. G. Blakeley, A. W. Charleson, et al., Recommendations for the design and construction of base isolated structures, Bull. N.Z. Nat. Soc. Earthquake Eng., 12(2), 136–157 (1979), with permission of NZNSEE.*]

load limiter is a flexural or torsional member with various possible configurations and is made of steel, lead, or other materials. When deformed plastically, it works as a damper to minimize the response and limit the amount of horizontal movement. In Fig. 4-27, the horizontal mounting and the load limiter are built as one piece, with the limiter being the central lead cylinder which resists flexure.

The concept of base isolation is best suited for low-rise, rigid structures located on hard ground. On the other hand, it is not suitable for use in a high-rise building since the overturning moment is too large. It is often the case that the natural period of a rigid building coincides with the predominant period of hard ground. Should this occur, resonance phenomena can be avoided by adopting the isolation concept, hence by protracting the natural period.

The dynamic response of a building with an isolated base tends to be complicated in comparison with that of a building without base isolation. It is therefore necessary to check dynamic response by an inelastic time-history response analysis using a design earthquake or earthquakes suited to the particular building site under consideration (Blakeley, Charleson, et al., 1979).

4.5.4.2 Energy-Dissipation Devices   Efforts have been made to devise artificial dampers to consume a portion of the ground-motion energy that is introduced into a building structure. Some are now in use. Oil dampers are not often used since they require frequent and troublesome maintenance.

Shear deformation of sandwich materials of high viscosity is sometimes utilized as an energy-dissipation device. A more common and simpler device utilizes the plastic flexural or torsional deformation of steels (Skinner, Kelly, and Heine, 1973). Such dampers are inserted in the middle of bracings, at wall-to-wall joints, or at the borders of a wall and a surrounding frame. Sometimes, steel plates with slits are placed between the wall and the surrounding frame as in Fig. 4-25. In this case, it is important to avoid excessive degradation and brittle fracture of the element under large repeated deformation.

4.6 Earthquake-Resistant Design of Structural Components and Systems

4.6.1 Introduction

Designing the connections and details of a structure to be earthquake-resistant is almost as important as checking the structure's overall dynamic behavior. If the strength and ductility of the connections are not adequate and if the details are not properly designed, the structure as a whole is not likely to display effective seismic performance. In this section, critical factors to be considered in the seismic design of connections and structural details will be discussed in relation to the structural characteristics described in Chap. 3.

In many design codes, the equivalent static earthquake force is combined with dead and live loads to evaluate the seismic safety of a structure. For instance, ATC-3 prescribes the following equations for the condition of earthquake loading (ATC-3, 1978):

$$U = 1.2D + 1.0L + 1.0S \pm 1.0E \tag{4-16}$$

or $\quad U = 0.8D \pm 1.0E \tag{4-17}$

in which U = the design load combination and $D, L, S,$ and E = the dead, live, snow, and earthquake loads, respectively.

Design should be made to satisfy the following equation:

$$\phi U_n \geqq U \tag{4-18}$$

in which U_n = the nominal strength and ϕ = the strength-reduction factor. ϕ is a coefficient to take into account possible reductions in the strength of members. Major causes of such reductions are error in design computation, variation of material properties, construction uncertainty, and dimensional error. Equation (4-18) follows the strength design concepts by which the ultimate strength of sections and members is designed. Some design codes adopt the allowable-stress design concept. In such codes, the allowable stress used to check seismic safety is frequently taken to be 30 to 50 percent larger than the allowable stress with which safety is evaluated under gravity load conditions.

4.6.2 Monolithic Reinforced-Concrete Structures

4.6.2.1 Introduction As discussed in Sec. 3.3.1, the monolithic reinforced-concrete structure is one of the most popular earthquake-resistant structural systems in the world. Various improvements in code provisions and in design pracice for this structural system have been achieved by study of the results of previous earthquakes, and earthquake damage of monolithic reinforced-concrete structures has decreased significantly in recent years. Important design rules for earthquake-resistant reinforced-concrete structures are similar to those for steel structures. They may be summarized as:

1. Ductility and large energy-dissipation capacity (with less deterioration in stiffness) must be provided.

2. Beams should yield before columns.

3. Flexural failure should precede shear failure.

4. Connections should be stronger than the members which frame into them.

The ACI Code (ACI Committee 318, 1983a, 1983b) has adopted the following equations for the combination of static design forces in place of Eqs. (4-16) and (4-17):

$$U = 1.4D + 1.7L \qquad (4\text{-}19)$$

$$U = 0.75(1.4D + 1.7L + 1.87E) \qquad (4\text{-}20)$$

$$U = 0.9D + 1.43E \qquad (4\text{-}21)$$

The ACI strength-reduction factor ϕ is given as:

Flexure without axial thrust	$\phi = 0.90$
Axial tension; axial tension with flexure	$\phi = 0.90$
Axial compression; axial compression with flexure	
Member with spiral reinforcement	$\phi = 0.75$
Member with other transverse reinforcement	$\phi = 0.70$
Shear and torsion	$\phi = 0.85$

ATC-3 basically uses the same values for ϕ, with some parts adopting slightly more conservative values.

4.6.2.2 Selection of Materials   The concrete in reinforced-concrete structures should not be of low strength. It is also advisable not to use brittle aggregates. ATC-3 recommends normal concrete with a strength of at least 20.7 MPa.

Yield stress and ductility are the important design properties for steel. Standards of the American Society for Testing and Materials (ASTM) and others stipulate lower limits for yield stress. Tension and bending tests provide useful information on steel ductility. Cold-worked steel is known to be less ductile, and steel having a significantly higher yield stress than nominal values also tends to show less ductile behavior. Use of such steel therefore is not recommended in most construction.

4.6.2.3 Beams   A member subjected to an axial force of $0.1 fc'A_g$ or less can be treated as a beam (A_g = the gross area of the section). The flexural capacity of a beam can be checked by computing M_u from Eq. (3-1) and then inserting M_u (nominal strength) into Eq. (4-18).

In accordance with the basic design philosophy that flexural failure should occur prior to shear failure, the required shear strength of a beam can be computed in the following manner. Let us suppose that a hinge is generated at each end of a beam (as shown in Fig. 4-28). We then obtain

$$V_u = \frac{M_{u1} + M_{u2}}{\ell} + V_g \tag{4-22}$$

in which M_{u1} and M_{u2} are the bending capacities of the plastic hinges at the ends and V_g is the shear stress caused by the vertically distributed load W. According to Eq. (4-18), the required nominal shear strength is given as $V_n \geq V_u/0.85$, and the required amount of reinforcement is calculated from Eq. (3-23).

Since the real f_y is normally greater than the standard value and steel stress is likely to fall into the strain-hardening range, $1.25f_y$ for normal steel ($1.4f_y$ for high-tension steel) is used in place of f_y to calculate the nominal strengths M_{u1} and M_{u2} in Eq. (4-22). Furthermore, the effect of slab reinforcement on M_u should be considered in this nominal-strength computation.

A very conservative approach is to ignore the contribution of concrete to shear resistance in a hinge zone ($v_c = 0$; Park and Paulay, 1980). As discussed in Sec. 3.3.4.2, diagonal reinforcement is effective in preventing brittle shear failure. Such reinforcement is also useful in that it increases shear and bending capacities proportionally.

To ensure sufficient ductility in beams, good design details are necessary. Critical requirements for design details are as follows (ACI Committee 318, 1983a, 1983b) (see Fig. 4-29):

1. At least a pair of longitudinal reinforcing bars is required for both top and bottom reinforcement. The steel ratio ρ should not be less than $1.38/f_y$ (MPa) and should not exceed 0.025. Positive bending strength should not be smaller than 50 percent of negative bending strength. Neither the negative- nor the positive-moment strength at any section

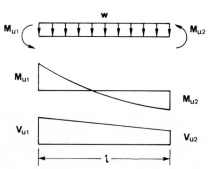

Figure 4-28 Calculation of required shear force.

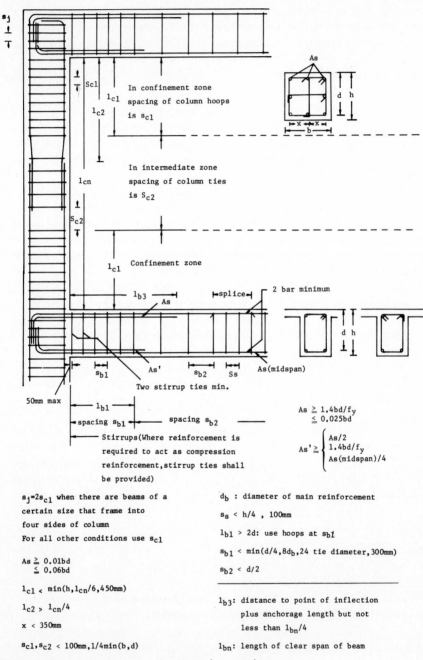

In confinement zone spacing of column hoops is s_{c1}

In intermediate zone spacing of column ties is S_{c2}

Confinement zone

splice

2 bar minimum

As(midspan)

Two stirrup ties min.

50mm max

spacing s_{b1} spacing s_{b2}

Stirrups(Where reinforcement is required to act as compression reinforcement,stirrup ties shall be provided)

$As \geq 1.4bd/f_y$
$\leq 0.025bd$

$As' \geq \begin{cases} As/2 \\ 1.4bd/f_y \\ As(midspan)/4 \end{cases}$

$s_j=2s_{c1}$ when there are beams of a certain size that frame into four sides of column
For all other conditions use s_{c1}

$As \geq 0.01bd$
$\leq 0.06bd$

$l_{c1} < min(h,l_{cn}/6,450mm)$

$l_{c2} > l_{cn}/4$

$x < 350mm$

$s_{c1},s_{c2} < 100mm,1/4min(b,d)$

d_b : diameter of main reinforcement

$s_s < h/4$, $100mm$

$l_{b1} > 2d$: use hoops at s_{b1}

$s_{b1} < min(d/4,8d_b,24$ tie diameter,$300mm)$

$s_{b2} < d/2$

l_{b3}: distance to point of inflection plus anchorage length but not less than $l_{bn}/4$

l_{bn}: length of clear span of beam

Figure 4-29   Details of beams, columns, and connections.

along the member length should be less than one-fourth of the maximum moment strength provided at the face of either point.

2. Web reinforcement should be arranged so that it can resist the sum of the shear produced by the factored gravity load and the shear sustained when a hinge occurs at each edge of the beam.

3. Hoops should be provided over a length equal to twice the member depth measured from the face of the supporting member toward midspan or over lengths equal to twice the member depth on both sides of a section where flexural yielding may occur. The first hoop should be located not more than 50 mm from the supporting member. Maximum spacing of the hoops should be less than $d/4$, 8 times the diameter of the smallest longitudinal bars, 24 times the diameter of the hoop bars, and 300 mm.

Where hoops are not required, stirrups should be spaced at no more than $d/2$ throughout the length of the member. Lap splices should not all be made at one section but rather be dispersed over the length. Lap splices of flexural reinforcement should be permitted only if hoop or spiral reinforcement is provided over the lap length.

The maximum spacing of transverse reinforcement enclosing the lapped bars should not exceed $d/4$ or 100 mm. Lap splices should be avoided within and near a joint or in any region where high stress is expected, particularly in a potential-yield hinge zone. In such high-stress regions, concrete is likely to crack, and this often results in insufficient stress transfer between the spliced reinforcing bars.

ACI 318-83 stipulates the following equations for the development length ℓ_d of deformed bars in tension whose diameters are 31 mm or smaller and which have $f_y \leq 414$ MPa:

$$\ell_d = 0.019\, A_b\, f_y / \sqrt{f_c'} \tag{4-23}$$

where $A_b =$ the area of an individual bar. Further, it is specified that ℓ_d should be at least 1.4 times longer than the value given by Eq. (4-23) for top reinforcement and $(2 - 414/f_y)$ times longer for bars with $F_y > 414$ MPa.

4.6.2.4 Columns Nominal strength N_n is computed by Eq. (3-11) for a concentrically loaded column and by Eqs. (3-12) and (3-13) for a column subjected to combined bending and axial force. The strength should satisfy the condition $N_n \geq N_u/\phi$.

As in the shear design of beams, the designed shear force V_u for a column should be taken as the shear sustained by the column when plastic hinges occur at the column or adjacent beams. The required nominal shear strength is given by $V_n \geq U_u/\phi$, and the amount of required shear reinforcement can be computed according to Eq. (3-23). Spacing of the shear reinforcement should not exceed $d/2$. As discussed in Secs. 3.3.4.2

and 3.3.4.3, diagonal reinforcement in columns is also effective in preventing brittle shear failure without increasing the number of hoop ties.

Besides the strength check, the ACI Code stipulates various requirements to ensure sufficient ductility of columns (ACI Committee 318, 1983a):

1. The steel ratio of longitudinal reinforcing bars should be between 0.01 and 0.06.

2. At a beam-column connection, the sum of flexural strengths of the columns under the factored axial-load condition should not be smaller than the sum of flexural strengths of the beams. If this condition is not satisfied, the column should be provided with transverse reinforcement as shown in item 3 over their full height.

3. If the maximum factored axial force is not greater than $A_g f_c'/10$ (A_g = the gross area of the section), column detail requirements can simply be the same as those for beams. If it is greater, however, ductility is likely to be reduced. In this case, confinement reinforcement should be provided over a length ℓ_0 from each joint face and on both sides of any section where flexural yielding may occur. The length ℓ_0 should not be less than h (column width), one-sixth of the column clear height, or 457 mm, whichever is the smallest. In columns supporting discontinued stiff members as walls and trusses, confinement reinforcement should be placed throughout their length.

If spiral hoops are used, the amount of reinforcement should not be smaller than either (a) the amount needed to ensure that no strength reduction will occur even when the cover concrete falls off:

$$\rho_s = 0.45 \left(\frac{A_g}{A_c} - 1 \right) \frac{f_c'}{f_y} \tag{4-24}$$

or (b) the amount needed to ensure adequate ductility:

$$\rho_s = 0.12 \frac{f_c'}{f_y} \tag{4-25}$$

whichever is smaller. A_c is the area of the core.

If rectangular hoops are used, transverse reinforcement should be larger than that of spiral hoops. The total cross-sectional area of rectangular-hoop reinforcement should not be less than that given by Eqs. (4-26) and (4-27):

$$A_{sh} = 0.3(sh_c \ f_c'/f_{yh})[(A_g/A_{ch}) - 1)] \tag{4-26}$$

$$A_{sh} = 0.12 \ sh_c \ f_c'/f_{yh} \tag{4-27}$$

where A_c = area of the core of the spirally reinforced compression member

A_{ch} = cross-sectional area of a member measured out to out of the transverse reinforcement

A_g = gross area of the section

A_{sh} = total cross-sectional area of the transverse reinforcement (including crossties) within spacing s and perpendicular to the dimension h_c

h_c = cross-sectional dimension of the column core

s = spacing of the transverse reinforcement

f_{yh} = specified yield strength of the transverse reinforcement (See Fig. 4-30.)

If the design strength of the member core satisfies the requirements of the specified loading combinations, Eqs. (4-24) and (4-26) need not be satisfied.

4. When longitudinal reinforcement is lap-spliced, splices are permitted only within the center half of the member length and should be proportioned as tension splices. If welding or mechanical splices are used, they may be employed for splicing the reinforcement at any section, provided no more than alternate longitudinal bars are spliced at a section and the distance between splices is 600 mm or more along the longitudinal axis of the reinforcement.

5. Transverse reinforcement should be spaced at distances not exceeding one-fourth of the minimum member dimension and 100 mm.

6. Spacing of crossties of legs of overlapping hoops should not be more than 350 mm.

4.6.2.5 Connections   A beam-to-column connection must be designed so that it does not fail before failure of the members framing into the

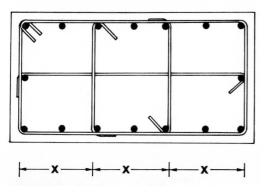

Figure 4-30   Confinement reinforcement.

connection. Three types of connection failure are common: shear failure in the panel zone, anchorage failure of longitudinal bars, and bond failure in longitudinal bars passing through the panel. As seen in the expression of Eq. (3-30), panel shear stress transferred from the member ends increases as flexural strength of those members increases, i.e., the steel ratio for the longitudinal bars becomes larger, and as shear stress in the members becomes smaller, i.e., the members become more slender. If shear stress exceeds a certain value, shear reinforcement must be provided in the panel zone.

Figure 4-31 shows a combination of stresses applied to the connection panel. Here, the longitudinal bars in tension are assumed to be yielding. To allow for a possible increase over the nominal yield value, a stress of α (stress multiplier) times the nominal value is used as the design yield stress. The ACI Code (ACI 318-83) and others (ACI-ASCE, 1976; Park and Paulay, 1980) recommend a value of 1.25 for α.

Referring to Fig. 4-31, the force applied to the upper face of the panel is

$$V_p = \alpha A_{s1} f_y + \alpha A_{s1} f'_s + \alpha C_{c1} - V_{col}$$
$$= (A_{s1} + A_{s2})\alpha f_y - V_{col} \tag{4-28}$$

Shear strength is given as the sum of the shear resisted by concrete and shear reinforcement [Eqs. (3-17) and (3-23)].

ACI-ASCE (1976) assumes that $V_c = 0$ if the column is subjected to tensile force. Park and Paulay (1980) are even more conservative and neglect the contribution of the concrete ($V_c = 0$). By using the nominal shear strength thus obtained, the required strength [Eq. (4-27)] and the reduction factor ϕ [Eq. (4-18)] can be checked. As discussed in Sec. 3.3.6, the shear resisted by the concrete (V_c) is smaller in T- or L-type connections than in cruciform connections. The AIJ Code for composite structures (AIJ, 1975b) reduces V_c for T- and L-type connections, respectively, to two-thirds and one-third of the V_c value for cruciform connections.

In connections where the width of the beam is smaller than the width

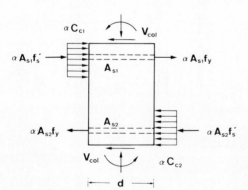

Figure 4-31 Forces on connection panel at ultimate state.

of the column, stress transfer from the beam and column to the panel becomes less effective, and as a result the shear resistance of the panel is smaller. A good practice therefore is to make the dimensions of the beam and column as close as possible.

According to the ACI Code, the same amount of transverse reinforcement as in columns is needed in joints, but transverse reinforcement can be halved if four beams extend from the panel and the beam width is not less than three-fourths of the column width (confined joint).

ACI 318-83 gives $20\sqrt{f_c'}A_i$ as the shear strength of a confined joint which has a specified amount of transverse reinforcement, where A_i is the minimum cross-sectional area of the joint in a plane parallel to the axis of the reinforcement generating the shear force. In this case the effect of transverse reinforcement on the shear strength of the joint is neglected.

ACI-ASCE (1976) stipulates an increase in panel reinforcement when column axial force is large, as in the design of columns. In this case, confinement reinforcement can be assumed to act as shear reinforcement.

The ACI Code (ACI 318-83, 1983) specifies that the development length ℓ_{dh} for a bar with a standard 90° hook should not be less than $8\,d_b$, 150 mm, and the length required by Eq. (4-29) (see Fig. 4-32):

$$\ell_{dh} = 0.18\,f_y\,d_b/\sqrt{f_c'} \tag{4-29}$$

For longitudinal beam bars which pass through a connection panel, Park and Paulay (1980) suggest that the ratio of the bar diameter to the column depth should not be greater than 0.04 to avoid bond slip of the bars. However, the strength reduction of an end cross section of a beam is

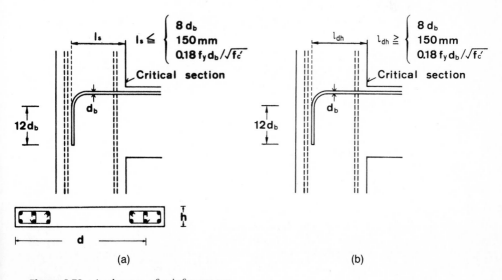

Figure 4-32   Anchorage of reinforcement.

usually not large when slip occurs, as discussed in Sec. 3.3.6, and this bar-size limit may be too stringent.

4.6.2.6 Shear Walls   As in the design of other structural elements, high strength, good energy-dissipation capacity, and minimal stiffness degradation are essential if shear walls are to be earthquake-resistant.

Cantilever Shear Walls. A shear wall must resist a combination of overturning moment, vertical load, and shear force. Even when shear walls are incorporated in a moment-resisting space frame in which a reduction in design horizontal force is allowed, shear reinforcement in the shear walls should be designed on the basis of the unreduced design horizontal force (ACI Committee 318, 1983).

To prevent a sharp reduction in flexural strength upon concrete cracking, vertical reinforcing bars should be placed near each edge of a shear wall so that they can resist the design moment (Fig. 4-33). Placing such reinforcing bars in the wall faces, instead of distributing them evenly over the width, increases both the flexural capacity and the ductility of the wall (Park and Paulay, 1980). Flanges placed at the faces are also effective (Fig. 4-33).

When a shear wall is designed to fail in the flexural mode, confinement reinforcement should be arranged in the region from the compression face to half of the depth of the neutral axis as shown in Fig. 4-33 (Paulay, 1980). When the axial force applied to a shear wall exceeds $0.4\phi P_b$, boundary columns should be introduced and designed to carry the sum of the axial forces caused by the vertical load and the overturning moment. Plate buckling may occur if shear walls or flange walls are subjected to high compressive force, and the thickness of such walls should be selected with care (Park and Paulay, 1980).

For a cantilever whose height-to-length ratio is large (often called a *slender wall*), horizontal reinforcement should be placed so that shear failure does not precede flexural failure. The amount of reinforcement

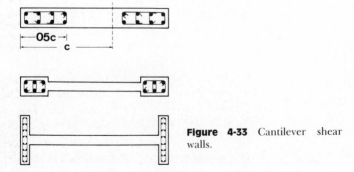

Figure 4-33 Cantilever shear walls.

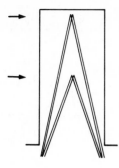

Figure 4-34 Application of diagonal reinforcement to shear walls.

should be obtained by the equation described in Sec. 3.3.5.2. In addition, the horizontal reinforcing bars should be anchored to the longitudinal boundary members.

In a shear wall which is liable to sliding failure as shown in Fig. 3-32c, it is useful to distribute the vertical reinforcing bars over the width of the shear wall. Diagonal reinforcement shown in Fig. 4-34 is also effective in dissipating energy when a wall undergoes sliding failure. Care should also be taken that the overturning moment transferred from a shear wall will not damage the foundation supporting this wall.

The horizontal shear reinforcement ratio ρ_h should not be smaller than 0.0025, and the vertical shear reinforcement ratio ρ_v should not be smaller than

$$\rho_v = 0.0025 + 0.5(2.5 - h_w/\ell_w)(\rho_h - 0.0025) \tag{4-30}$$

or 0.0025, whichever is greater, but need not be greater than ρ_h (ACI Committee 318, 1983). In the vicinity of a potential plastic hinge, the spacing of transverse ties and shear reinforcing bars should be decreased.

The amount of vertical reinforcement needed to resist sliding in a construction joint should be determined in the following way. Assuming that 75 percent of the axial force acting in the wall is effective in providing frictional force between the sliding faces and using the shear friction theory to take into account the effect of reinforcement on the friction, we obtain the following equation:

$$V_j = \phi(A_{vf}f_y + 0.75N) \tag{4-31}$$

in which A_{vf} is the total amount of vertical reinforcement, including the reinforcement allocated for flexural resistance. For lightweight concrete, ϕ is taken to be 60 percent of the value specified for normal-weight concrete. Horizontal construction joints should be thoroughly cleaned and roughened.

Shear Walls with Opening. Diagonal reinforcement is useful for coupling beams, as shown below in Fig. 3-36*a*, and wall columns, as shown below in Fig. 3-36*b*.

If the compressive stress at an open face exceeds $0.2f_c$, a boundary element should be extended to a point where the compressive force is reduced to $0.15f_c$ (ACI Committee 318, 1983).

Low-Rise Shear Walls. For a low-rise shear wall having no boundary element or for such a wall framed only by weak elements, the amount of horizontal and vertical reinforcement should be determined on the basis of the shear strength computed by Eq. (3-26). If a low-rise shear wall is clamped by a stiff, strong frame, Eq. (3-28) should be used instead.

4.6.2.7 Floor Slabs   In addition to their primary function as vertical load-resisting elements, floor slabs act as elements which distribute earthquake force into the vertical structural elements. This action is often referred to as *diaphragm action.* The force-resisting mechanism in floor slabs under earthquake conditions is very complex because of the interaction between out-of-plane and in-plane forces acting on them. It is known, however, that these two types of forces can be treated independently in most floor-slab designs (Aoyama and Yoshimura, 1980; Karadogan, Huang, et al., 1980; Nakashima, Lu, et al., 1980).

In-plane shear is produced in floor slabs when the earthquake force acting on one floor is transferred to stiff earthquake-resisting elements such as shear walls and is not significant in most building structures. On some occasions, however, in-plane characteristics of floor slabs should be incorporated into earthquake-resistant design. When there is an opening in a floor slab, some reinforcement may be needed at the corners. For a building with a staggered wall system, floor slabs sustain the earthquake force directly, and their in-plane strength should be examined carefully. If a building layout includes a wing, as in Fig. 4-14, the in-plane stress in a floor slab may be very high in the proximity of the corner. In this case, the in-plane strength should be checked.

4.6.3 Precast-Concrete Structures

4.6.3.1 Introduction   Precast concrete (PC) is one of the most popular structural systems throughout the world. In Europe over 25 percent of building construction uses this system, while in the U.S.S.R. the PC system is said to exceed 80 percent of building construction. In the United States, the PC system is used in more than two-thirds of housing production, much of which occurs in seismic regions (Tall Building Committee 21E, 1979; Hawkins, 1981). The PC system has poor overall integrity because of its connections, and, as a matter of fact, making the connections sufficiently strong and ductile is extremely difficult. Accordingly, much damage of PC structures has occurred in past earthquakes. To build earthquake-resistant PC structures, the design must follow the rules used for reinforced-concrete structures. In addition, connections should be carefully designed to be strong as well as ductile, while site connections should be located in low-stress regions. For the design of connections in PC structures, ATC-3 (1978) stipulates the use of 0.5 as the reduction coefficient ϕ. This is significantly lower than the value specified for reinforced-concrete structures. Reference ATC-8 (1981) contains an SOA report (State-of-the-Art Report) on precast concrete.

The PC system is classified into two types: the frame system and the panel system. The frame system is further divided into two groups: the linear system and the frame subassemblage system. In the linear system, precast column and beam elements are assembled at the construction site (Fig. 4-35a). In the frame subassemblage system, components such as T, cruciform, H, π, and hollow panel frames are assembled at the site (Fig. 4-35b). It is very difficult in a linear system to ensure sufficient strength and ductility at beam-to-column connections, and such systems are not suitable as moment-resisting structures. The linear system, therefore, is

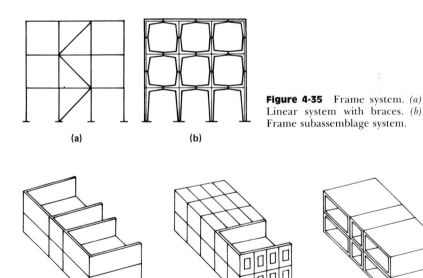

(a) (b)

Figure 4-35 Frame system. *(a)* Linear system with braces. *(b)* Frame subassemblage system.

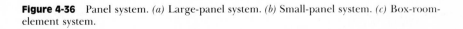

(a) (b) (c)

Figure 4-36 Panel system. *(a)* Large-panel system. *(b)* Small-panel system. *(c)* Box-room-element system.

often combined with cast-in-place shear walls or steel braces. In the frame subassemblage system, selection of joint locations is more flexible, and connections are usually located in low-stress regions such as the inflection points in columns.

The panel system has many variations including the small-panel system, the large-panel system, and the box-room system (Fig. 4-36). For all of these, the strength and ductility of the connections are the critical design considerations (Mueller, 1981).

Precast columns, beams, and panels are sometimes designed so that their ends or edges extend into cast-in-place joints. By this means, better integrity can be achieved between precast elements.

4.6.3.2 Connections   Figure 4-37 shows examples of joints used in the linear system. Good connection details are needed to ensure sufficient strength and ductility of joints where stress level is high (Shinagawa, 1973). Joints in the subassemblage system are relatively easy to design because they are usually located in the vicinity of inflection points of precast beams and columns. Figure 4-38 shows an example of such a joint. A steel plate is shop-welded at the end of each PC column, and the two plates are field-welded.

Joints for wall-to-wall connections may be vertical or horizontal. Vertical joints are not stressed under conditions of vertical load but are

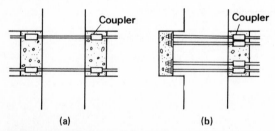

(a)   (b)

Figure 4-37   Beam-to-column connection. *(a)* Reinforced concrete. *(b)* Prestressed concrete.

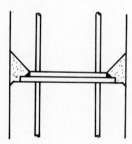

Figure 4-38   Column-to-column joint.

subjected to shear when horizontal force is applied to the wall. The connections therefore should be reinforced to resist shear.

Figure 4-39 illustrates examples of vertical connections. In type *a*, reinforcing bars extend from the walls and are lapped at the joint, so that force transfer depends on the bond of the bars. In type *b*, the bars are welded at the joint. In the vertical connection in *c*, shear is resisted by shear keys and shear failure occurs as a result of concrete crushing, sliding failure, or diagonal-tension failure.

A horizontal joint resists a combination of compressive force due to vertical load, shear force due to horizontal force, and, in some cases, tensile force due to in-plane bending of the wall.

A floor-to-floor joint is subjected to a combined compressive (or tensile) force and shear. The compressive (or tensile) force results from slab bending, whereas shear is caused by the diaphragm action of the slab. Figure 4-40 shows examples of floor-to-floor joints. In type *a*, the reinforcing bars which extend from each slab panel are field-welded at the joint. In type *b*, a pair of steel plates, each shop-welded to the anchoring bar, are jointed by welding a splice plate. In detail *b*, both compressive (or tensile) force and shear transfers can be achieved.

Figure 4-41 shows a typical American platform-type connection.

4.6.4 Prestressed-Concrete Structures

4.6.4.1 Earthquake-Resistant Design of Structures   As discussed in Sec. 3.2.7, the earthquake response of prestressed-concrete structures is significantly higher than that of reinforced-concrete structures. This is

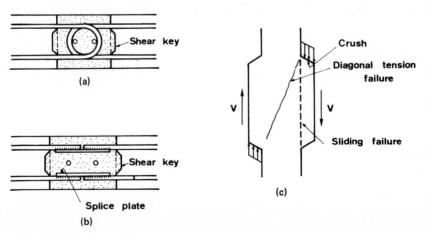

Figure 4-39 Wall-to-wall joint. *(a)* Section of shear key with lapped bars. *(b)* Section of shear key with welded bars. *(c)* Side view of shear key.

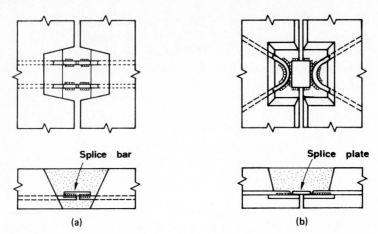

Figure 4-40   Floor-to-floor joint. *(a)* Lap-welded joint. *(b)* Plate-welded joint.

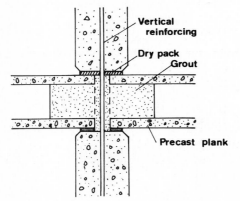

Figure 4-41   Platform connection.

reflected in the New Zealand code, which stipulates design earthquake forces 20 percent higher than those normally specified for reinforced-concrete structures (Park and Paulay, 1980).

As in the case of reinforced-concrete structures, prestressed-concrete structures should be designed carefully so that the structure possesses sufficient strength and ductility. In the structure, beams should yield prior to columns, and connections should not fail before attached beams and columns. It is not good practice to have the prestressing cables unbonded in major force-resisting elements. Prestressing steel and ordinary reinforcing bars should be used together to ensure ductility. A mixed structure in which reinforced-concrete elements are combined with prestressed-concrete elements is known to be more earthquake-resistant than a prestressed-concrete structure.

4.6.4.2 Beams   To guarantee sufficient ductility, the steel ratio $w_p = \rho_p f_{ps}/f'_c$ (discussed in Sec. 3.3.8.3) should not be greater than 0.2, and the flexural cracking load should not be larger than flexural ultimate strength. Shear reinforcement should be provided so that flexure failure precedes shear failure.

The region of a potential plastic hinge is taken to be $2h$ with h as the beam depth, and stirrups should be included in this region to ensure concrete confinement, prevent buckling of reinforcing bars, and act as shear reinforcement. Stirrup spacing should not be greater than 150 mm, $d/4$, or six longitudinal bar diameters, whichever is the smallest (Park, 1980).

4.6.4.3 Columns   Ultimate flexural strength should not be smaller than the flexural cracking moment, and shear reinforcement should be used to ensure that flexure failure precedes shear failure. Transverse reinforcing bars should be provided in any potential plastic-hinge region, with spacing not greater than one-fifth of column width, six longitudinal bar diameters, or 200 mm, whichever is the smallest (Park, 1980).

4.6.4.4 Connections   Since a joint core is subjected to high diagonal tension under earthquake loading, prestressing steel should not be anchored at the joint core. For a prestressed beam framed into an exterior column, prestressing steel may well be anchored in a concrete stub attached on the outer surface of the column (Fig. 4-37). Shear design of a joint core should follow the design specification for reinforced-concrete joints. Prestressing steel placed at middepth of a beam is effective in resisting the diagonal tension of the joint core.

4.6.5 Steel Structures

4.6.5.1 Introduction   Structural planning of steel structures should be in accordance with the basic conditions discussed in Sec. 4.5. Members and connections should be designed so as to guarantee high strength, ductility, and energy-dissipation capacity (or equivalently minimum degradation). Limits must therefore be placed on the width-to-thickness ratio, slenderness ratio, axial force, and spacing of lateral braces in order to avoid premature instability. The selection of materials and the detailed design of connections should be carried out with care to prevent brittle fracture. As already noted, beams should yield prior to columns, and the strength of a connection should be larger than the strength of the beams and columns which frame into the connection. The steel itself should be (1) homogeneous, with moderate values for both upper and lower yield stresses; (2) ductile and feasible to weld; and (3) without laminations. ATC-3 suggests the use of A36-75, A441-75, A500-76, A501-76, A572-76, and A588-75 as specified by the ASTM.

ATC-3 sets ϕ at 0.9 for the design of connections which are to resist the forces transmitted from the attached members. The nominal strength to be transferred to a connection is determined by multiplying the allowable forces (which are computed on the basis of Part 1 of AISC specifications) by 1.7. For special moment frames, the nominal strength should be the ultimate strength calculated according to Part 2 of the AISC specifications. Table 4-2 shows several examples of the allowable maximum width-to-thickness ratio. In the Japanese code, the factor D_s in Table 4-1 varies with the value of this width-to-thickness ratio. Unbraced structures are ductile and possess large energy-dissipation capacity but tend to deform greatly. To prevent serious damage to nonstructural elements under small- to medium-size earthquakes or strong wind, the drift should be checked (Sec. 4.2.7). The major causes of this large deformation are:

1. Flexural and shear deformations of beams and columns

2. Shear deformation of connections

3. Overall structural bending produced by column elongation and contraction

4. Overall structural rotation produced by foundation uplift and settlement

To make the design effective, it is useful to identify the dominant causes of the total deformation.

Braced frames can resist larger amounts of horizontal forces and have reduced lateral deflection and thus reduced $P\text{-}\Delta$ effect. As discussed in Sec. 4.5.3.2, however, if bracing is arranged concentrically in the structure, it often becomes very difficult to prevent foundation uplift. In this case, bracing should be distributed uniformly throughout the structure. The hysteretic behavior of braces is usually of the degrading type and therefore involves little energy dissipation because of the alternating buckling and plastic elongation under load reversals. To allow for this degradation

TABLE 4-2 Limitation of Width-Thickness Ratio of Plate Elements

	F_y	A-36	A-441
Flanges of rolled wide-flange shape ($b_f/2t_f$)		8.5	7.0
Flanges of box sections (b/t)		31.7	26.9
Webs (d/t)	$P/P_y \leq 0.27$	$68.7-96.1P/P_y$	$58.3-81.6P/P_y$
	$P/P_y > 0.27$	42.8	36.3

effect in design, some codes require an increase in the design earthquake force of braced frames by a constant factor over that specified for unbraced frames. For instance, the UBC stipulates that the earthquake force specified for a ductile frame system be increased by 20 percent for a dual bracing system in which at least 25 percent of the total earthquake force is carried by special moment frames. In the Japanese code, earthquake force is increased in accordance with the force resisted by the bracing relative to the frame, with a 50 percent increase as the upper limit.

4.6.5.2 Beams   In a potential plastic-hinge region, the width-to-thickness ratio of the beam should be kept small, and the lateral braces should be spaced with a small pitch to ensure sufficient rotation capacity of the beam. As discussed in Sec. 3.4.3.2, local or lateral buckling is likely to occur at a lower load under load reversals than under monotonic loading. Nevertheless, the provisions given in Part 2 of the AISC specifications (AISC, 1978) are recommended for use even though they have been based primarily on studies of monotonically loaded beams, the reasons being:

1. The restraining effect of floor slabs on beam strength is considered to be significant.

2. The rotation capacity required for beams under earthquake loading is still very unclear.

The AIJ specification also adopts this concept so that the width-to-thickness ratio of beams is limited according to Table 4-2 and the spacing of lateral braces is limited according to Eq. (3-38). Outside of plastic-hinge regions, beams need only resist external forces (ductility is not required), and therefore a larger spacing is allowable. It should be noted, however, that in most cases the locations of plastic hinges are unpredictable.

4.6.5.3 Columns   Strength reduction under load reversals is very significant if a column is subjected to large axial force. Axial force therefore should be kept smaller than 60 percent of yield axial force. As in beam design, in a potential plastic-hinge region the width-to-thickness ratio and the spacing of lateral braces should be limited, respectively, according to Table 4-2 and Eq. (3-38). In other regions the width-to-thickness ratio need only be not greater than the ratio specified for elastic design, while lateral braces should be spaced to provide the required strength.

When a column is slender and is subjected to large axial force, column ductility is small, as shown in Fig. 3-83. According to the AIJ guide for plastic design (AIJ, 1975a), the slenderness ratio of axially loaded columns (made of steel equivalent to A36) should satisfy the following condition:

$$K\ell/r \leqq 150 \qquad \text{for} \qquad N/N_y < 0.15 \qquad (4\text{-}32)$$

$$\frac{N}{N_y} + \frac{K\ell/r}{120} \leqq 1.0 \qquad \text{for} \qquad N/N_y \geqq 0.15 \qquad (4\text{-}33)$$

in which N = the axial compressive force, N_y = the yield axial force, $K\ell$ = the effective buckling length, and r = the radius of gyration of the column.

The AIJ guide for high-rise structures (AIJ, 1973) recommends that the slenderness ratio be limited to an even smaller value than that specified by Eqs. (4-32) and (4-33). In this guide, it is assumed that a rotation capacity of 3 is sufficient to ensure ductility in high-rise structures.

4.6.5.4 Bracing Bar braces are often used in small structures. They should be designed so that their ends do not fracture before the bar yields. On the other hand, braces used in most building structures (except small ones) have a shaped or tube cross section. ATC-3 stipulates that the compressive strength of a brace should not be smaller than half of the tensile yield force of the brace. Even following this provision, we find that deflection in the direction perpendicular to the brace longitudinal axis can be very significantly large once the brace buckles, and damage of some nonstructural elements may then be unavoidable. It is recommended (AIJ, 1973) that $K\ell/r$ values for braces should not exceed 30 in high-rise building structures.

A design rule for braces is that the connections should not fracture prior to yielding of the brace. To achieve this goal, axial yield force (i.e., the cross-sectional area of the brace multiplied by yield stress) should be smaller than the strength of the connections. Because of alternating buckling and plastic elongation, hysteretic behavior of braces is often of the degrading type. In the United States and Japan, eccentric bracing systems as shown in Fig. 3-73 have been used for the construction of high-rise buildings (Merovich, Nicoletti, and Hartle, 1982).

Local buckling sometimes occurs after a brace buckles; buckling of a built-up brace may also be followed by cracking in the brace. Bracing in important buildings therefore should consist of wide flanges or tubes rather than built-ups. In addition, care must be taken so that no stress concentration is generated in the connections. If a brace is placed eccentrically to the surrounding columns and beams, a torsional moment is induced in those members. To avoid this moment, eccentricity should be minimized.

4.6.5.5 Connections   A general rule for connection design is that the strength of a connection should not be smaller than the strength of the ends of the member that is framed into the connection. However, if the rotation capacity of a connection is verified as large by experiment or analysis, the design connection force can be reduced to a value equal to the member-end force at which the member receives twice the deflection computed on the basis of the external design force (ATC-3, 1978). As for the strength in a panel zone, not only must the panel be capable of resisting forces much higher than yield strength, but it must also have large deformation capacity. It may be allowable to use as the panel's nominal strength a force higher than yield strength (Sec. 3.4.6.1).

To ensure sufficient strength and ductility at a joint, welding should be used so that maximum member strength can be transferred safely to the panel. In addition, details should be designed to avoid stress concentration or lamellar tearing.

The Japanese code specifies for mild-steel structures that the ratio of maximum strength of connections to yield strength of members should be at least 1.2 or 1.3 in beam-to-column joints and splices of beams or columns in flexure and also for joints of braces in tension.

Equations (3-44) and (3-45) should be used to check for crippling in a column web owing to the compressive force transferred from the beam flange, and stiffeners should be provided if necessary. A safety check of the web for the tensile force transferred from the beam flange should be made according to Eq. (3-46). If the column's axial force exceeds half of its yield axial force, the shear stress applied to the panel should be determined on the basis of combined axial and shear forces. If a doubler plate is used to strengthen the panel, it must be welded directly to the column web.

4.6.6 Composite Structures

4.6.6.1 Introduction   The characteristics of concrete-encased structures as compared with reinforced-concrete and steel structures can be described as follows: Concrete-encased structures are more ductile than reinforced-concrete structures, yet they are stiffer and less prone to buckling than steel structures. On the other hand, they are more expensive than reinforced-concrete structures and more difficult to construct than steel structures.

Concrete-filled tubes have various advantages over reinforced-concrete columns. They are more ductile and possess larger strength because of the restraining effect of the steel tube on the infilled concrete, and furthermore they do not need formwork for construction. They are also superior to steel columns, in that they are more effective against buckling and require less fireproofing. The connection details, however, are much more complicated than those used in either reinforced-concrete or steel structures.

The flexural strength of a composite beam in positive bending is very high because of the contribution of the compressive-concrete slab. In negative bending, however, the slab has little to contribute, and its flexural strength is not much greater than that of the steel beam. To ensure the diaphragm action of the floor slabs, the slabs should be rigidly connected to the steel beam. For this purpose, shear connectors are often used. As in reinforced-concrete design, the design of composite structures should follow these basic rules:

1. Beam yielding precedes column yielding.

2. Flexural failure precedes shear failure.

3. Connection strength is greater than the strength of members framed into the connection.

There is no provision with respect to the strength-reduction factor ϕ for composite cross sections. In brief, use of the mean value of the reduction factors specified for reinforced-concrete and steel sections is recommended. The design strength of the section is then the nominal strength of a composite section, computed by the method of superposed strength, multiplied by the mean reduction factor. The recommended reduction factors to be used for composite sections are $\phi = 0.9$ for flexural and tension members, $\phi = 0.83$ for compression members with spiral hoops, $\phi = 0.80$ for other types of members, and $\phi = 0.88$ for connections. The reduction factor for concrete-filled tubes may be taken to be equal to that for steel members; i.e., $\phi = 0.9$. A design check of encased members against shear should be made separately for steel and reinforced-concrete portions, utilizing the respective reduction factors.

As discussed in Sec. 3.5.1, in a mixed system of construction the joints between two different types of members are often the weakest part of the structure. Such joints require very careful design.

4.6.6.2 Beams and Columns   The ratio of the area of the major steel component to that of the reinforcing bars can be chosen arbitrarily in designing an encased member. It is recommended that the major area be as large as possible, so that good ductility in shear can be achieved. There is another good reason to recommend this practice: with increasing longitudinal reinforcement, an increasing amount of shear reinforcement is required to ensure flexural failure in the reinforced-concrete portion of the member.

Nominal flexural strength of encased beams and columns can be computed by means of the superposed-strength method discussed in Sec. 3.5.2.1 (AIJ, 1975*b*).

A shear check for an encased member should be made separately for the steel and reinforced-concrete portions rather than for the composite member. The primary reason for this procedure is that in an encased member the steel section usually has a small flange and a thick web, while the reinforced concrete has a considerable amount of longitudinal reinforcement with little hoop reinforcement. In such cases hysteretic behavior is very poor. Required shear strength of the reinforced-concrete portion should be computed by Eq. (4-22) with M_{u1} and M_{u2} as the ultimate moments at the ends of the portion. In applying this equation, the total shear caused by the gravity load V_g is assumed to be resisted by the steel and reinforced-concrete portions in proportion to their respective steel ratio. Equation (4-22) should also be used to check the steel portion against shear. Because the steel portion is still ductile even after it sustains shear yielding, M_{u1} and M_{u2} may be taken as the steel end moments when the design external force is applied rather than the full plastic moments.

The flexural design of concrete-filled tube columns can simply follow that of encased columns. As for the shear design, a check of the concrete portion is not needed since the infilled concrete does not exhibit shear failure (as discussed in Sec. 3.5.3). The check for the steel-tube portion follows the procedure specified for encased columns.

The nominal shear strength of a wide-flange section in an encased member should be calculated as $(F_y/\sqrt{3})td_w$ (see Sec. 3.4.3.1). For a tube section with a cross-sectional area A_t, the strength should be calculated as $(F_y/\sqrt{3})A_t/2$.

To ensure sufficient ductility for an encased column, axial force applied to the column should not exceed the force given by Eq. (3-53).

Design details for the reinforced-concrete parts of encased beams and columns should follow those specified for reinforced-concrete beams and columns.

4.6.6.3 Connections

As in other types of structures, connections in composite structures should be designed so that they do not fail before the yielding of the ends of the members that frame into those connections.

The required shear strength V_p of the panel in an encased beam-to-column joint can be obtained by employing Eq. (3-30). In this equation, the superposed ultimate flexural strength should be used for M_{b1} and M_{b2}, and the shear force equilibrating those moments for V_{c1} and V_{c2}. Panel resistance should be evaluated by Eq. (3-56), and on the basis of this resistance the necessary web thickness of the steel panel and the amount of shear reinforcement should be determined. As shear reinforcement is difficult to arrange in a shear panel, the panel should not be reinforced more than is needed for confinement.

Stiffeners should be welded to the steel in a panel if it is calculated to be necessary (see Sec. 4.6.5.5). Their shape should be determined carefully so that they do not interfere with the placing of concrete. Longitudinal reinforcing bars in a beam should not be curtailed in the middle of the panel but must extend through to the other side of the panel. If a steel web extends perpendicularly to the bars, it should be pierced to allow the bars to extend to the other side.

As discussed in this section, joint design for encased structures is very difficult. Great care must be taken in the design process even when a practicable form of reinforcement can be achieved in the joints.

4.6.7 Masonry Structures

4.6.7.1 Introduction   Masonry structures are characterized by high stiffness and great weight. As a result, energy dissipation into the ground is large; nevertheless, earthquake response is high because of the short period. Maximum acceleration response can easily rise to as much as 3 times ground acceleration. For instance, in an earthquake as large as the El Centro earthquake (north-south component, 1940), acceleration response can reach about 1 g. This acceleration is significantly higher than the seismic coefficients specified in various design codes (Priestley, 1980). As in other types of structures, economical design can be achieved in masonry structures if energy dissipation can be assumed to occur as a result of ductile behavior. It is difficult to achieve ductility in masonry structures as compared with reinforced-concrete structures. Nevertheless, it is possible if steel is combined effectively with masonry and details are designed with care (Tall Building Committee 27, 1979). In addition to the design rules described in Sec. 4.6.2, important rules for the seismic design of masonry structures are as follows (AIJ, 1970, 1979):

1. A structure should be an assembly of box-type units. Long or high walls are not recommended.

2. Openings should be minimized, and the total length of the walls should extend as far as possible in both horizontal directions. The AIJ Code introduces a term *wall ratio*, which is defined as the total wall length divided by the floor area and is calculated separately for each horizontal direction. In each direction, the wall ratio should not be smaller than 150 mm/m^2 on the average and 210 mm/m^2 for the upper three stories. The aspect ratio of walls and piers should be as small as possible.

3. Reinforced-concrete beams should be placed at the top of all masonry walls or at each floor or roof level. These beams are effective in preventing out-of-plane failure of the walls.

4. Openings of walls and shear walls should not be eccentrically located on the structural plan. Since eccentricity in the plan of the center of stiffness relative to the center of gravity causes torsional moment of the structure, such eccentricity should be minimized.

5. Walls in the upper story should be placed directly on top of the walls in the story below.

6. Shear reinforcement should be provided in walls to ensure their ductile behavior.

7. Floor-to-wall or wall-to-wall connections should be secured with reinforcing bars, and these connections should be grouted by mortar or concrete to maintain their integrity.

8. High, slender masonry structures such as chimneys should be designed for increased design earthquake forces.

9. Stiff, strong continuous footings should be used for the foundation.

4.6.7.2 Member Design   Member design of masonry structures generally follows the allowable-stress design of reinforced-concrete members. As in the design of steel or timber structures, ATC-3 specifies that the ultimate strength of a masonry member should be calculated as its allowable stress multiplied by a certain factor. This strength then should be used to check the basic requirements described in Sec. 4.6-1.

The ultimate strength of a masonry member is thus given as the allowable stress multiplied by 2.5, while ϕ's values to be used in Eq. (4-18) are as follows:

Compression or combined bending and compression and
bearing stress $\phi = 1.0$

    Reinforcement except shear $\phi = 0.8$

    Shear reinforcement $\phi = 0.6$

    Shear carried by masonry $\phi = 0.4$

For a compressed-masonry member, ultimate strength is 2.5 times as large as its allowable stress and the ϕ value adopted for this member is 1. For flexural reinforcement in a masonry member, ultimate strength is 2 times as large as allowable stress. Instead of using a different multiplier, i.e., 2.0, allowable stress is still multiplied by 2.5, the same value as for a compressive member, and to compensate for the discrepancy ϕ is set at $2.0/2.5 = 0.8$. Other ϕ values are chosen by using the same concept.

The design of masonry structures as stipulated in the ACI Committee Report (ACI Committee 531, 1970) and ATC-3 (1978) is summarized below. Allowable flexural moment of a masonry member is calculated on the basis of the assumptions that a plane section remains plane after deformation occurs and that the allowable tensile stress of masonry is zero.

The allowable axial stress of a masonry column is given by

$$P_a = A_g(0.20^* f'_m + 0.65 p_g f_s)\left[1 - \left(\frac{h}{40t}\right)^3\right] \tag{4-34}$$

The allowable axial stress of a masonry wall is given by

$$F_a = 0.225^* f'_m \left[1 - \left(\frac{h}{40t}\right)^3\right] \tag{4-35}$$

Note that ATC-3 specifies 0.18 instead of 0.20* in Eq. (4-34) and 0.20 instead of 0.225* in (Eq. 4-35). Both of these equations include the effect of strength reduction caused by buckling. Symbols used in Eqs. (4-34) and (4-35) are as follows:

A_g = gross area of column
f'_m = masonry compressive strength
p_g = ratio of effective cross-sectional area of vertical reinforcement to A_g
f_s = allowable stress in reinforcement
h = unsupported height of column or wall
t = total depth of wall

The compressive force in a masonry structure subjected to combined bending and axial force should satisfy the following condition:

$$\frac{f_a}{F_a} + \frac{f_m}{F_b} \leq 1 \tag{4-36}$$

where f_a = calculated axial stress
f_m = calculated flexural compressive stress
F_b = permissible flexural stress

Further, if a masonry wall is subjected to in-plane seismic force, total compressive force should not exceed F_a given by Eq. (4-35) (ATC-3, 1978).

The relationship between shear force V and shear stress v acting on a flexural member is

$$V = vbjd \tag{4-37}$$

in which b = width of the compression flange of the flexural member

j = ratio of the distance between the centroid of compression and the centroid of tension reinforcing bars to depth d (j may be taken as 0.8 or determined by a strain-compatibility analysis)

d = effective depth of the section

When external shear force is assumed to be resisted only be shear reinforcement with a cross-sectional area A_v, the allowable shear V carried by the reinforcement is

$$V = A_v f_v j d/s \tag{4-38}$$

in which A_v = total area of web reinforcement within a distance of s

s = spacing of the stirrups

f_v = tensile strength of web reinforcement

According to the ACI Code, the unit bond stress u acting on the reinforcing bars is

$$u = \frac{V}{\Sigma_0 j d} \tag{4-39}$$

in which Σ_0 = the sum of the perimeters of all bars. Also, the necessary development length for the bars (ATC-3) is

$$\ell_g = 0.038 d_b^2 f_y / \sqrt{f_g} \tag{4-40}$$

where d_b = diameter of the smaller bar spliced

f_y = specified strength

f_g = strength of the mortar or grout

4.6.7.3 Minimum Reinforcement and Details

The maximum amount of reinforcement and the details to be followed in designing masonry walls, columns, and beams have been specified in many design codes. Important provisions for the seismic design of masonry members in ATC-3 will be mentioned here.

For a masonry shear wall, the reinforcement ratio should not be smaller than 0.0015 in both the vertical and the horizontal directions. If the reinforcement has sufficient capacity to resist external shear force, the ratio can be reduced accordingly, but it should not be smaller than 0.0007 in each direction. In addition, the sum of the ratios in the horizontal and vertical directions should not be smaller than 0.002. Spacing of vertical and horizontal reinforcement should not be greater than one-third of the length and the height, respectively, or 810 mm.

A tie anchorage needs a 135° turn plus at least six tie diameters or a 100-mm extension, whichever is greater.

The diameter of the longitudinal reinforcement in a masonry wall should not be smaller than 12 mm, and it generally should not exceed 30 mm. Furthermore, the reinforcement ratio for a column should be between 0.005 and 0.04.

A column subjected to large axial compression or tension under earthquake loading, such as a boundary column attached to a masonry wall, must be securely reinforced by stirrups or crossties over its entire length. For a column not resisting high axial force, stirrups or ties are still needed in the end regions. Spacing of these ties should not be greater than 16 bar diameters or 200 mm, whichever is smaller. Ties are needed even in less critical regions, and spacing should not be greater than 16 bar diameters, 48 tie diameters, the least column diameter, or 440 mm, whichever is smallest.

The lap length of a deformed bar should not be smaller than the length calculated by Eq. (4-40), with 0.061 instead of 0.038 in this equation.

4.6.7.4 Nonstructural Walls   As will be discussed in Sec. 4.7, installation of nonstructural walls often changes the dynamic behavior of a structure or causes stress concentration in some part of the structure. Design of the structure should therefore include the possible effects of such walls. Alternatively, they can be isolated from the structure.

The design detail shown below in Fig. 4-43 can be used to attach a nonstructural wall to a frame and provide lateral support to the wall. If such a detail is difficult to fabricate, the wall should be placed as a cantilever fastened to the floor. In this case, the joint should be reinforced to sustain the maximum possible bending moment. This design detail should be used also for partition walls.

Veneer glued to the surface of a wall in a timber structure often falls off owing to its failure to follow the wall deflection. If veneer is attached to a wall, wall stiffness should therefore be commensurate with the in-plane stiffness of the veneer (Glogau, 1974).

4.6.8 Timber Structures

In the design of timber structures, the relationship between allowable stress and ultimate strength differs significantly from those in steel or reinforced-concrete structures because timber design allows for the effect of creep. In ATC-3, ultimate strength is taken to be twice allowable stress. For ϕ's ATC-3 thus specifies $\phi = 1.0$ for bending, bearing compression, and tension; and $\phi = 0.75$ for shear on diaphragms and shear walls. In this provision, design details and allowable-stress values are recommended for various types of diaphragms. A limit is also set on the type of diaphragms that can be used as shear walls, according to the seismic zone and the importance of the structure. On the basis of observed damage in previous earthquakes, important rules for the ductile design of timber structures are as follows (Iizuka, 1980; Evans 1973):

1. The plan should not be large.

2. Walls should be arranged as symmetrically as possible to minimize torsional moment applied to the structure.

3. Shear walls and columns should be supported on reinforced-concrete footings, and reinforcing bars in the walls and columns must be securely anchored in the footings. Steel plates are to be used to connect walls or columns in the first and second stories.

4. Steel devices such as metal corner plates and toothed steel connections should be used for connections.

5. Large openings in walls are not desirable.

6. Horizontal diaphragms should be arranged to prevent relative horizontal deflection between vertical walls and columns.

7. A diagonal brace should be framed into the adjoining vertical members and nailed. A hole drilled for nailing should be slightly smaller than the nail diameter so that the brace is not split at the nail hole. The resistance of nails to withdrawal forces should be neglected when the nails are placed parallel to the grain.

8. Metal laths or chicken wire should be used for plaster-wall cladding.

9. Roofing should be as light as possible.

10. Care should be taken in the installation of window glass to ensure safety against structural deformation.

4.7 Design of Nonstructural Elements

4.7.1 Introduction

Architectural components and mechanical components are classified as *nonstructural elements*. Architectural components include nonbearing walls, veneers, roofing units, partitions, stairs and shafts, and ceilings. Mechanical components include boilers, communication systems, electrical bus ducts, and machinery. The seismic behavior of nonstructural elements has not been adequately studied, and effective design specifications are practically nonexistent. The seismic safety of nonstructural elements is therefore largely dependent on the skill and conscience of manufacturers and installers.

Significant damage of nonstructural elements has been reported after strong earthquakes such as the San Fernando earthquake (1971), the Managua earthquake (1972), the Guatemala earthquake (1976), and the Miyagiken-Oki earthquake (1978). In the San Fernando and Guatemala earthquakes, damage of nonstructural elements accounted for more than 50 and 70 percent, respectively, of the total damage cost. After the Miyagiken-Oki earthquake, over 60 percent of the nonstructural bearing walls of a high-rise apartment building were severely damaged even though there was very little structural damage (Tiedeman, 1980; AIJ, 1980).

As the use of nonstructural elements in modern buildings is great, structural engineers must bear in mind that failure of these elements can destroy human-safety systems such as those relating to fire resistance and evacuation.

Provisions for the seismic design of nonstructural elements have been developed on the basis of past experience and have appeared in codes and specifications (such as ATC-3-06, 1978) since the early 1970s.

In general, nonstructural elements fail because of either excessive inertial forces applied to them or excessive deflection caused by deformation of the structural system. The inertial force acting on a nonstructural element can be predicted if the response of the structure is known at the floor level where the nonstructural element is installed. Maximum shear acting on a flexible element may be significantly greater than shear acting on a similar rigid element. On the other hand, if a nonstructural element is subjected to forced deflections, it must be capable of deforming without failure in accordance with the structural deformation. As a design alternative, the element may be detached from the structure so that deformation of the structural system does not affect deformation or force in the element.

As is also very true in structural design, structural engineers must combine human safety with economy in the design of nonstructural elements. Falling ceilings, window glass, and exterior walls may kill people in the vicinity. Collapse of stairways and damage of exit doors may prevent the escape of people from the building. Emergency lighting and exit signs should not malfunction during and after earthquake disturbances. Important buildings such as hospitals, emergency disaster centers, and fire stations must remain functioning immediately after earthquakes. Clearly emergency systems and important building structures must possess higher than normal safety.

Expensive nonstructural elements, as well as dangerous elements like those in nuclear power plants, in which damage could cause disasters, need to be designed with great care. To determine the design force applied to such a nonstructural element, the floor response of the structure where the element is installed is first computed. On the basis of this response, the response of the element can then be determined. Other nonstructural elements may not need such a sophisticated (but cumbersome) procedure for determining the design force. An equivalent static lateral force, including a significant safety margin, may be used to design the elements. In currently available design provisions, this equivalent static force is usually prescribed.

Although the effects of nonstructural elements on the behavior of a structural system are usually considered to be secondary, interactive behavior between structural and nonstructural elements has been reported as the cause of structural failure. Whenever unfavorable interactive behavior is expected, nonstructural elements may well be detached from the structural system or analyzed by taking the interaction into consideration.

4.7.2 Dynamic Forces Applied to Nonstructural Elements

When a rigid nonstructural element is tightly clamped on the floor of a structure, the response of the element is identical with the floor response. The magnification factor, defined as the ratio of the element response to the floor response, is therefore unity. When a rigid nonstructural element is installed by a flexible connecting device on the floor of a structure, the element response is greater than the floor response. Such behavior can be represented by a one-mass system with damping. One mass has as many as six degrees of freedom, but the system can usually be simplified to one with a single degree of freedom. Vibrational characteristics of the part of the structural system where the nonstructural element is placed, often represented as the floor acceleration response spectra, can be found by applying time-history response analysis to the structural system. If the connecting device is ductile, the magnification factor derived from elastic response analysis may be relaxed according to ductility.

The floor response spectra for designing nonstructural components were proposed by Skinner (1964). If such a spectrum and the natural period of a nonstructural element are known, the design force of the element can be computed with great ease.

A long, flexible nonstructural element such as a piping system or a cable tray cannot be simplified to a single-degree-of-freedom system. An analysis of a multiple-degrees-of-freedom system is therefore required, with the floor response as input. Equivalent static analysis is usually adequate for the design of such a system unless its failure is considered to cause crucial damage in the structural system.

4.7.3 Equivalent Static Analysis

Various design codes stipulate the equivalent static force to be sustained by nonstructural elements. Design procedures prescribed in a code (ATC-3, 1978) will be discussed below.

4.7.3.1 Architectural Elements   The equivalent static force to be applied to an architectural element is obtained from the following equation:

$$F_p = A_v C_c P W_c \tag{4-41}$$

where F_p = seismic force applied to an element
C_c = seismic coefficient for elements
W_c = weight of the component
A_v = seismic coefficient representing the effective peak velocity-related acceleration
P = performance-criteria factor

A_v is specified according to the location where the structure is built; the same values should be used as for structures. The seismic coefficient is set in accordance with the type of element, and values between 0.9 and 3.0 are specified. The performance-criteria factor indicates the importance of the nonstructural element; it is set in accordance with the type of element and the type of structure in which the element is placed. Values between 0.5 and 1.5 are specified. In the design procedure, structures are classified into three groups, referred to as seismic-exposure groups. In the classification, the importance of the structure, the effect of damage to the structure on the public, and human response are considered. Group III contains building structures which must be operational for rehabilitation purposes immediately after disasters. Buildings open to the public are classified in Group II, whereas others are in Group I.

4.7.3.2 Mechanical and Electrical Elements   An equivalent static force to be applied to a mechanical or electrical element is computed by

$$F_p = A_v C_c P a_c a_x W_c \tag{4-42}$$

where C_c = seismic coefficient for an element (values between 0.67 and 2.0 are specified)
a_c = amplification factor related to the response of a system or element as affected by the type of attachment
a_x = amplification factor at level x related to the variation of the response in the height of the building

The amplification factor a_x is derived from the equation

$$a_x = 1.0 + h_x/h_n \tag{4-43}$$

where h_x = height above the base to level x
h_n = height above the base to level n
n = number of stories

4.7.4 Interaction Effects
on Architectural Nonstructural Elements

When a nonstructural wall is tightly clamped in a structural frame, the wall is forced to deform in a compatible manner with the frame. The wall fails if it is forced by the frame to deform beyond its allowable limit. To avoid such failure, the wall may be uncoupled from the frame. Figure 4-42 shows an example of uncoupled-wall systems. In Fig. 4-42*b* the wall is fastened to the frame at four corners by an attachment which allows the wall to slide freely in the wall plane but strongly resist out-of-plane deformation. The clearance distance between the wall and the frame needs to be determined by taking into account possible drift of the frame.

To prevent water leakage in exterior walls and to satisfy acoustic and fire-resistance requirements in interior walls, the clearance should be padded by fillers as in Fig. 4-43.

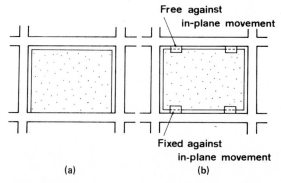

Figure 4-42 Isolation of nonstructural walls from frames.
(a) Three-corner isolation. *(b)* Isolation by steel plates.

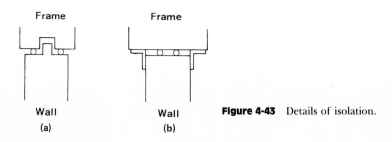

Figure 4-43 Details of isolation.

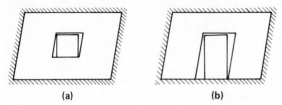

Figure 4-44 Clearances between windows or doors and walls. *(a)* Window. *(b)* Door.

Breakage of window glass is very dangerous because falling pieces can injure people below. If an expected maximum frame deformation is considered to be small, the glass can be fixed by soft putty. If it is large, clearance must be provided between the window sash and surrounding walls and frames as shown in Fig. 4-44a. When walls surrounding a doorway are subject to large deformation, the doorway may become jammed. Proper clearance must be provided as shown in Fig. 4-44b.

Ceiling failure causes critical damage in many instances. Since suspended ceilings often fall during earthquake disturbances, connections with the suspending members must be properly designed. Care also must be taken so that ceilings do not hit surrounding walls in the course of their horizontal movement. Furthermore, design precautions must be taken to prevent ceiling finishes and lighting fixtures from falling to the floor.

4.7.5 Effects of Nonstructural Elements on Structural Systems

In the normal practice of structural design, nonstructural elements are not taken into account. Completed structures, however, contain various nonstructural elements such as claddings and exterior and partition walls which influence structural behavior under earthquakes. Structural engineers cannot ignore the influence in some situations. The influence is found to be small if flexible nonstructural elements are added to a stiff structural system. In the reversed situation, for example, when exterior or partition walls made of masonry or block concrete are installed in a frame, the influence must be great.

Various effects of nonstructural elements on structural behavior are conceivable:

1. The natural period of the structural system may be shortened, resulting in a different input level to the system.

2. Distribution of story shear in columns may change, and some columns may sustain more force than that assumed in the original design.

3. An unsymmetrical arrangement of nonstructural walls may cause significant torsion in the system.

4. Local force may be concentrated if nonstructural walls are rearranged nonuniformly in height.

Structural engineers have two approaches to mitigating such unfavorable conditions. First, nonstructural elements can be uncoupled from the structural system as described in Sec. 4.7.4. This approach is used almost exclusively in designing tall buildings. In the second approach, nonstructural elements can be treated as structural members and their characteristics taken into account in the design. This makes insulation of water, noise, or heat more feasible than in the first approach. In general, however, stiff and brittle nonstructural elements such as wing walls are more vulnerable than structural members because of their lower deformation capacity. Furthermore, hysteretic behavior of the structural system is very complex when such nonstructural elements are included. The complexity often results in a poor understanding of the true response of the structure.

Figure 4-45a illustrates examples of the first approach. By providing a slit between the spandrel beam and the column, the column is expected to behave in a ductile manner. As in Fig. 4-45b, the same treatment is possible in wing walls so that adjoined beams would not fail by shear mode.

4.7.6 Design Details for Mechanical and Electrical Elements

4.7.6.1 Equipment   Rigid equipment such as a transformer of a boiler may move horizontally or fall down by rocking motion during earthquake vibration. If pipes and electric service hoses are connected to such equipment, the connections will be damaged. For small and less important

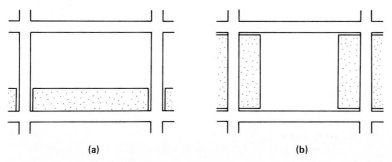

| (a) | (b) |

Figure 4-45 Isolation of nonstructural walls from the structural frame. *(a)* Vertical isolation. *(b)* Horizontal isolation.

equipment, the equipment base and floor can be painted on their interfaces to increase friction. To improve seismic resistance, the equipment should be clamped to the floor with bolts. The amount of anchoring required to resist horizontal force should be determined in accordance with the shape of the equipment: height of the center of gravity and height-to-width ratio. Design care should also be taken so that the anchor can resist possible vertical vibration.

When equipment is suspended from the ceiling, cross bracing as shown in Fig. 4-46 is effective in preventing it from moving.

4.7.6.2 Equipment with Vibration Isolation   Much damage has been reported on mechanical systems set on vibration-isolation devices, which are used primarily for acoustic control. Such systems have to satisfy two incompatible, conflicting requirements: the requirement to be able to dissipate vibration (need for flexibility) versus the requirement to be able to sustain earthquake loads (need for rigidity). The failure of isolation devices causes mechanical systems to fall onto the floor.

To make a floor-mounted vibration-isolation device seismically effective, the equipment base should be clamped to the floor by bolts. As an alternative method, a restraining device may be placed next to the equipment, restricting both its vertical and its horizontal movement (Fig. 4-47). For light equipment suspended by hanger rods, cross bracing performs well in restricting vibration of the equipment as in Fig. 4-48, while a restraining frame is more effective for heavy equipment. Design details are depicted in Ayres and Sun (1972).

4.7.6.3 Piping and Air-Conditioning Systems   Little damage has been reported on piping systems after earthquakes. As shown in Fig. 4-49, piping systems and ducts may well be braced by rods in both longitudinal and transverse directions. Relative movement of the floor to which a piping system is screwed can distort pipes. Particularly when a pipe extends over an expansion joint, it must be flexibly connected to the floor.

4.7.6.4 Light Fixtures   Three commonly used types of light fixtures are (1) recessed fixtures, (2) surface-mounted fixtures, and (3) suspended (pendant) fixtures. Minor damage has been caused by inadequate fixtures in systems 1 and 2. Contrary to popular opinion, damage of light fixtures sustained by ball joints was found to be significant. These fixtures can easily go into resonance.

Ball joints with a spring inside them can restrict the vertical motion of a light fixture. To reduce possible resonance, a damping device needs to be incorporated into the fixture (Ayres and Sun, 1972).

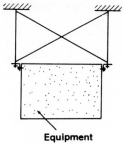

Figure 4-46   Cross bracing for equipment isolation.

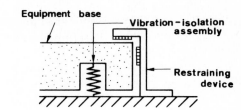

Figure 4-47   Stopper for a vibration-isolation system.

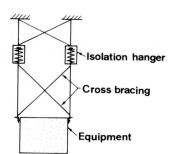

Figure 4-48   Bracing of vibration-isolation system.

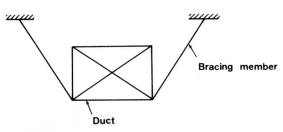

Figure 4-49   Sway bracing of ducts.

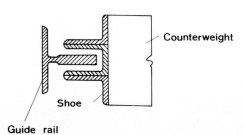

Figure 4-50   Safety shoe in an elevator counterweight.

4.7.6.5 Elevators Various surveys have shown that most damage to elevator systems occurs at counterweights and in their proximity. Many counterweights have become detached from guiderails, some smashing elevator cars. A strong guiderail and/or safety shoes (Fig. 4-50) would inhibit such damage (Ayres and Sun, 1972).

REFERENCES

ACI-ASCE Committee 352 (1976). Recommendations for design of beam-column joints in monolithic reinforced concrete structures, *J. Am. Concr. Inst.*, **73**(7), 375–393.

ACI Committee 318 (1983a). *Building Code Requirements for Reinforced Concrete (ACI 318-83)*, American Concrete Institute, Detroit.

—— (1983b). *Commentary on Building Code Requirements for Reinforced Concrete (ACI 318-83)*, American Concrete Institute, Detroit.

ACI Committee 531 (1970). Concrete masonry structures, design and construction, *J. Am. Concr. Inst.*, **67**(6), 442–460.

AIJ (Architectural Institute of Japan) (1970). *Design Essential in Earthquake Resistant Buildings*, Elsevier Publishing Company, Amsterdam.

—— (1973). *Design Guide for Tall Buildings*, AIJ, Tokyo (in Japanese).

—— (1975a). *Guide of the Plastic Design of Structures*, AIJ, Tokyo (in Japanese).

—— (1975b). *Design Standard for Composite Steel-Reinforced Concrete Structures*, AIJ, Tokyo (in Japanese).

—— (1979). *Design Standard for Special Concrete Structures and Commentary*, AIJ, Tokyo (in Japanese).

—— (1980). *Report on the Damage Due to 1978 Miyagiken-Oki Earthquake*, AIJ, Tokyo (in Japanese).

AISC (American Institute of Steel Construction) (1969). *AISC Specification for the Design, Fabrication and Erection of Structural Steel for Buildings, and Commentary*, AISC, New York.

—— (1978). *AISC Specification for the Design, Fabrication and Erection of Structural Steel for Buildings, and Commentary*, AISC, New York.

Aoyama, H., and M. Yoshimura (1980). Tests of RC shear walls subjected to bi-axial loading, *Proc. Seventh World Conf. Earthquake Eng.*, Istanbul, **7**, 511–518.

ATC-3 (Applied Technology Council) (1978). *Tentative Provisions for the Development of Seismic Regulations for Buildings (ATC 3-06)*, Nat. Bur. Stand. Spec. Publ. 510, Washington.

ATC-8 (Applied Technology Council) (1981). *Proceedings of the Workshop on Design of Prefabricated Concrete Buildings for Earthquake Loads*, Nat. Bur. Stand. Spec. Publ., Washington.

Ayres, J. M., and T.-Y. Sun (1972). Criteria building services and furnishings, in W. Wright, S. Kramer, and C. Culver (eds.), *Building Practice for Disaster Mitigation*, U.S. Department of Commerce, National Bureau of Standards, Washington, 253–285.

Blakeley, R. W. G., A. W. Charleson, H. C. Hitchcock, L. M. Megget, M. J. N. Priestley, R. D. Sharpe, and R. I. Skinner (1979). Recommendations for the design and construction of base isolated structures. *Bull. N. Z. Nat. Soc. Earthquake Eng.*, **12**(2), 136–157.

Blume, J. A. (1977). The SAM procedure for site-acceleration-magnitude relationships, *Proc. Sixth World Conf. Earthquake Eng.*, New Delhi, **1**, 416–422.

—— (1980). Distance partitioning in attenuation studies, *Proc. Seventh World Conf. Earthquake Eng.*, Istanbul, **2**, 403–410.

Bycroft, G. N. (1960). White noise representation of earthquakes, *J. Eng. Mech. Div., Am. Soc. Civ. Eng.*, **86**(EM-1).

Caspe, M. S. (1970). Earthquake isolation of multistory concrete structures, *J. Am. Concr. Inst.*, **67**, 923–933.

Cherry, S. (1974a). Earthquake ground motions: Measurement and characteristics, in J. Solnes (ed.), *Engineering Seismology and Earthquake Engineering*, Noordhoff International Publishing, Leiden, 87−124.

—— (1974b). Estimating underground motions from surface accelerograms. in J. Solnes (ed.), *Engineering Seismology and Earthquake Engineering*, Noordhoff International Publishing, Leiden, 125−150.

—— (1974c). Design input for seismic analysis, in J. Solnes (ed.), *Engineering Seismology and Earthquake Engineering*, Noordhoff International Publishing, Leiden, 151−162.

Chopra, A. K., and N. M. Newmark (1980). Analysis, in E. Rosenblueth (ed.), *Design of Earthquake Resistant Structures*, Pentech Press, London, 27−53.

Donovan, N. C. (1974). A statistical evaluation of strong motion data including the February 9, 1971, San Fernando earthquake, *Proc. Fifth World Conf. Earthquake Eng.*, Rome, **1,** 1252−1261.

Esteva, L., and E. Rosenblueth (1963). Espectros de temblores a distancias moderadas y grandes, *Proc. Chilean Conf. Seismol. Earthquake Eng.*, University of Chile, **1.**

Evans, F. W. (1973). Earthquake engineering for the smaller dwelling, *Proc. Fifth World Conf. Earthquake Eng.*, Rome, **2,** 3010−3013.

Glogau, O. A. (1974). Masonry performance in earthquakes, *Bull. N. Z. Nat. Soc. Earthquake Eng.*, **7**(4), 149−166.

Gutenberg, B., and C. F. Richter (1956). Earthquake magnitude, intensity, energy, and acceleration (second paper), *Bull. Seismol. Soc. Am.*, **46**(2), 105−146.

Hawkins, N. (1981). Precast concrete connections, *Proc. U.S./PRC Workshop Seismic Anal. Des. Reinf. Concr. Struct.*, Ann Arbor, Mich., 7−27.

Housner, G. W. (1965). Intensity of earthquake ground shaking near the causative fault, *Proc. Third World Conf. Earthquake Eng.*, Auckland, **1**(3), 94−111.

—— (1969). Engineering estimates of ground shaking and maximum earthquake magnitude, *Proc. Fourth World Conf. Earthquake Eng.*, Santiago, **1**(A-1), 1−13.

—— and P. C. Jennings (1964). Generation of artificial earthquake, *J. Eng. Mech. Div., Am. Soc. Civ. Eng.*, **90**(EM-1), 113−150.

—— and —— (1974). Problems in seismic zoning, in J. Solnes (ed.), *Engineering Seismology and Earthquake Engineering*, Noordhoff International Publishing, Leiden, 163−177.

IAEE (1980). *Earthquake Resistant Regulations: A World List, 1980*, compiled by International Association for Earthquake Engineering, Gakujutsu-Bunken-Fukyu-Kai (Association for Science Documents Information), Tokyo.

Iizuka, G. (1980). On the damage and caution to wooden houses by recent earthquakes in Japan, *Seventh World Conf. Earthquake Eng.*, Istanbul, **5,** 121−124.

Jennings, P. C., G. W. Housner, and N. C. Tsai (1969). Simulated earthquake motions for design purposes, *Proc. Fourth World Conf. Earthquake Eng.*, Santiago, **1**(A1), 145−160.

Kanai, K. (1960). An empirical formula for the spectrum of strong earthquake motion, *Proc. Second World Conf. Earthquake Eng.*, Tokyo, **3,** 1541−1561.

—— (1966). Improved empirical formula for the characteristics of strong earthquake motions, *Proc. Japan Earthquake Eng. Symp. Tokyo*, 1−4.

Karadogan, H., T. Huang, L.-W. Lu, and M. Nakashima (1980). Behavior of flat plate floor systems under in-plane seismic loading, *Proc. Seventh World Conf. Earthquake Eng.*, Istanbul, **9,** 9−16.

Lee, D. M., and I. C. Medland (1978a). Base isolation—A historical development, and the influence of higher mode responses, *Bull. N.Z. Nat. Soc. Earthquake Eng.*, **11**(4), 219−233.

—— and —— (1978b). Estimation of base isolated structures responses, *Bull. N.Z. Nat. Soc. Earthquake Eng.*, **11**(4), 234−244.

McGuire, R. K. (1977). Seismic design spectra and mapping procedures using hazard analysis based directly on oscillator response, *Earthquake Eng. Struct. Dyn.* **5**(3), 211−234.

Matsushita, K., and M. Izumi (1965). Deflection controlled elastic response of buildings and methods to decrease the effect of earthquake forces applied to buildings, *Proc. Third World Conf. Earthquake Eng.*, Auckland, **3,** 360−372.

Megget, L. M. (1978). Analysis and design of a base-isolated reinforced concrete frame building, *Bull. N.Z. Nat. Soc. Earthquake Eng.*, **11**(4), 245−254.

Merovich, A. T., J. P. Nicoletti, and E. Hartle (1982). Eccentric bracing in tall buildings, *J. Struct. Div., Am. Soc. Civ. Eng.*, **108**(ST-9), 2067−2080.

Mueller, P. (1981). Seismic behavior of precast walls, *Proc. U.S./PRC Workshop Seismic Anal. Des. Reinf. Concr. Struct.*, Ann Arbor, Mich., 7−27.

Muto, K. (1974). Earthquake resistant design of tall buildings in Japan, in J. Solnes (ed.), *Engineering Seismology and Earthquake Engineering*, Noordhoff International Publishing, Leiden, 203−245.

Nakashima, M., L.-W. Lu, T. Huang, and H. F. Karadogan (1980). Current research at Lehigh University on concrete floor systems under in-plane seismic loading, *Proc. Seventh World Conf. Earthquake Eng.*, Istanbul, **9**, 145−156.

Park, R. (1980). Partially prestressed concrete in seismic design of frames, *Proc. FIP Symp. Partial Prestressing Practical Constr. Prestressed Reinf. Concr.*, Bucharest, **1**, 104−117.

―――― and T. Pauley (1980). Concrete structures, in E. Rosenblueth (ed.), *Design of Earthquake Resistant Structures*, Pentech Press, London, 142−194.

Pauley T. (1980). Earthquake-resisting shear walls—New Zealand design, *J. Am. Concr. Inst.*, **80**(77), 144−152.

Petrovski, J., D. Jurukovski, and V. Simovski (1978). Dynamic response of building with isolation on rubber cushions, *Proc. Sixth Eur. Conf. Earthquake Eng.*, Dubrovnik, Yugoslavia, **3**, 29−36.

Priestley, M. J. N. (1980). Masonry, in E. Rosenblueth (ed.), *Design of Earthquake Resistant Structures*, Pentech Press, London, 195−222.

―――, R. L. Crosbe, and A. J. Carr (1977). Seismic forces in base-isolated masonry structures, *Bull. N.Z. Nat. Soc. Earthquake Eng.*, **10**(2), 55−68.

RCDERT (Research Committee for the Development of Earthquake-Resistant Technology) (1977). *General Report on the Development of Earthquake Resistant Technology*, Kokudo-Kaihatsu-Gijutsu Research Center, Tokyo (in Japanese).

Robinson, W. H., and A. G. Tucker (1977). A lead-rubber shear damper, *Bull. N.Z. Nat. Soc. Earthquake Eng.*, **10**(3), 151−153.

Schnabel, P., H. B. Seed, and J. Lysmer (1971). *Modification of Seismolograph Records for Effects of Local Conditions*, Rep. EERC-71-8, Earthquake Eng. Res. Center, College of Engineering, University of California, Berkeley.

Seed, H. B., I. M. Idriss, and F. W. Kiefer (1969). Characteristics of rock motions during earthquakes, *J. Soil Mech. Found. Div., Am. Soc. Civ. Eng.*, **95**(SM-5), 1199−1218.

Shinagawa, T. (1973). Connections of prefabricated reinforced concrete structures, *Concr. J.*, Japan National Council on Concrete, **11**(11), 75−81.

Skinner, R. I. (1964). Earthquake generated forces and movements in tall buildings, *Bull. Dep. Sci. Ind. Res.*, New Zealand, 166.

―――, J. M. Kelly, and A. J. Heine (1973). Energy absorption devices for earthquake resistant structures, *Proc. Fifth World Conf. Earthquake Eng.*, Rome, **2**, 2924−2933.

Tajimi, H. (1960). A statistical method of determining the maximum response of a building structure during an earthquake, *Proc. Second World Conf. Earthquake Eng.*, Tokyo, **2**, 781−797.

Tall Building Committee 15 (Committee 15 of the Council on Tall Buildings and Urban Habitat) (1979). Plastic analysis and design, in *Monograph on Planning and Design of Tall Buildings*, SB-3, American Society of Civil Engineers, New York, 137−238.

Tall Building Committee 16 (Committee 16 of the Council on Tall Buildings and Urban Habitat) (1979). Stability, in *Monograph on Planning and Design of Tall Buildings*, SB-4, American Society of Civil Engineers, New York, 239−342.

Tall Building Committee 21D (Committee 21D of the Council on Tall Buildings and Urban Habitat) (1979). Design of cast-in-place concrete, in *Monograph on Planning and Design of Tall Buildings*, CB-11, American Society of Civil Engineers, New York, 501−576.

Tall Building Committee 21E (Committee 21E of the Council on Tall Buildings and Urban Habitat) (1979). Design of structures with precast concrete elements, in *Monograph on Planning and Design of Tall Buildings*, CB-12, American Society of Civil Engineers, New York, 575−653.

Tall Building Committee 27 (Committee 27 of the Council on Tall Buildings and Urban Habitat) (1979). Design of masonry structures, in *Monograph on Planning and Design of Tall Buildings*, CB-13, American Society of Civil Engineers, New York, 655–708.

Tiedeman, H. (1980). A statistical evaluation of the importance of non-structural damage to buildings, *Proc. Seventh World Conf. Earthquake Eng.*, Istanbul, **6,** 617–624.

Toki, K. (1981). *Earthquake Resistant Analysis of Structures,* Gihodo Shuppan Co., Tokyo (in Japanese).

Trifunac, M. D., and A. G. Brady (1975). On the correlation of peak acceleration of strong motion with earthquake magnitude, epicentral distance and site conditions, *Proc. U.S. Nat. Conf. Earthquake Eng.*, Ann Arbor, Mich., 43–52.

Tyler, R. G. (1977). Dynamic tests on laminated rubber bearings, *Bull. N.Z. Nat. Soc. Earthquake Eng.*, **10**(3), 143–150.

UBC (Uniform Building Code) (1979). International Conference of Building Officials, Whittier, Calif.

5

ASEISMIC DESIGN OF FOUNDATIONS

5.1 Test of Soil Characteristics

In designing the foundations of a building structure, the designer must first determine the soil characteristics of the construction site. Borehole drilling and penetration tests are the two standard site tests. Data needed for the aseismic design of foundations can be obtained from the following tests.

5.1.1 Field Tests

Borehole drilling is usually made from ground level to a depth of about 2 to 4 times the width of the footings. The depth of drilling, however, depends very much upon the size and importance of the building.

Penetration-resistance testing is performed primarily for measuring the relative density and degree of compaction of the soil. Hollow-tube samplers and core penetrometers are used for this test. The tests usually are of two types: dynamic and static. The standard penetration-resistance test, popular in the United States and Japan, can also provide data to be used for judging potential liquefaction and for calculating the allowable bearing capacity of sandy ground.

The shear modulus of soil G can be estimated from a shear wave-velocity test. An explosive charge or a hammer is used to produce waves in the soil. The velocity is measured by applying the excitation at one borehole and measuring the velocity at another borehole or by applying an excitation on the ground and measuring the velocity at a borehole.

The fundamental period of soil is an important property for the earthquake-resistant design of structures. It can be estimated by a micro-tremor test or from a measurement of small earthquake disturbances.

5.1.2 Laboratory Tests

Soil samples collected from a construction site can be tested in a laboratory in order to find soil characteristics such as weight per unit volume, cohesion, internal friction angle, water content, liquid limit, plastic limit, void ratio, compression index, preconsolidated load, and sensitivities. Further, particle-size distribution and relative-density tests are useful to check potential soil liquefaction. The cyclic triaxial test is a useful means of estimating the damping ratio of the soil. In this test, first hydraulic pressure is applied to a cylindrical sample, and then reversed loading is applied to the cylinder in its longitudinal direction. From the stress-strain curve, the elastic modulus E is estimated. The shear modulus G can then be computed from E and the measured Poisson ratio. The damping ratio is computed by use of Eq. (2-165). The moduli E and G can also be determined by applying axial and torsional vibrations to the cylindrical sample.

5.1.3 Shear Modulus and Damping of Soils

5.1.3.1 Shear Modulus   Figure 5-1 illustrates typical shear-modulus versus shear-strain relationships for sand and saturated clay (Seed and Idriss, 1971). For both sand and clay, the shear modulus is reduced as the strain level increases. The shear modulus measured from the wave velocity corresponds to 10^{-5} to 10^{-4} of strain.

5.1.3.2 Damping

Material Damping. Figure 5-1 also shows the relation between damping and shear strain. Since in most cases the strain level experienced during earthquakes ranges from 10^{-3} to 10^{-1}, damping as high as 10 and 16 percent for clay and sand, respectively, can be expected during an earthquake.

Radiation Damping. There is no convenient method to measure radiation damping of soil at a construction site. According to the theoretical

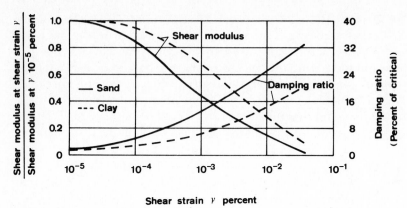

Figure 5-1 Average relationships of shear modulus and internal damping to shear strain for sands and saturated clays.

study of a circular foundation on a semi-infinite body, the radiation damping caused by translational motion of such a foundation is significantly larger than that caused by twisting or rocking motion (Whitman and Richart, 1967).

5.2 Dynamic Characteristics of Soils

5.2.1 Liquefaction of Saturated Sands

Liquefaction is defined as a phenomenon whereby a saturated sandy layer loses its shear strength owing to earthquake motion and behaves like liquid mud. Many earthquake reports refer to such liquefaction. In the Niigata earthquake of 1964, many buildings subsided, inclined, and even overturned as a result of liquefaction. Landslides induced by liquefaction damaged many dwellings in the Alaska earthquake of 1964 (Seed, 1970).

When a saturated sandy layer is subjected to reversed shear, pure-water pressure increases and in turn decreases the effective stress of the sand. When shear strength is reduced to zero, liquefaction occurs.

According to Ohsaki (1970), liquefaction is likely to occur under the following soil conditions:

1. The sandy layer is within 15 to 20 m of ground level and is not subjected to high overburden pressure.

2. The layer consists of uniform medium-size particles (Fig. 5-2).

3. The layer is saturated; i.e., it is below groundwater level.

4. The standard penetration-test value is below a certain level (Fig. 5-3; Seed and Idriss, 1971).

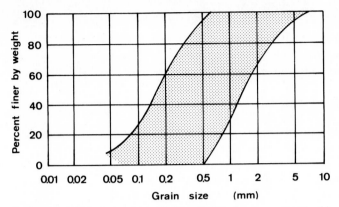

Figure 5-2 Critical zone for grain-size distribution curves by Ohsaki. [*Y. Ohsaki, Effects of sand compaction on liquefaction during the Tokachi-Oki earthquake,* Soil Found., *10(2), 112−128 (1970).*]

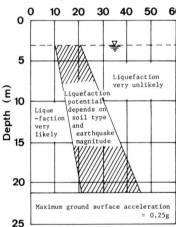

Figure 5-3 Liquefaction-potential evaluation chart by Seed. [*H. B. Seed and I. M. Idriss, Simplified procedure for evaluating soil liquefaction potential,* J. Soil Mech. Found. Div., Am. Soc. Civ. Eng., *97(SM-9), 1249−1273 (1971).*]

Seed and Idriss (1971) proposed an evaluation method for potential liquefaction based on the ratio of average shear stress to initial effective overburden pressure.

To lessen the possibility of liquefaction, the following measures can be considered:

1. Increase the relative density of the sands by compaction.
2. Replace the soil with soil which has less likelihood of liquefaction.
3. Install drainage equipment in the ground.
4. Drive piles to a layer which is less liable to liquefaction.

5.2.2 Settlement of Dry Sands

Loose, dry sands settle under earthquake vibration. The settlement is larger for sands with smaller relative density.

5.3 Design of Foundations

The ground condition at a construction site should be carefully examined for potential liquefaction, slope instability, or surface rupture due to faulting or lurching.

Critical factors associated with the design of foundations can be summarized as follows:

1. Input and output force characteristics of foundations. These are related both to the forces transmitted from the superstructure to the ground and to the forces transmitted to the piles, basements, and foundations from the ground.

2. Strength and deformation characteristics of foundations, piles, and piers.

3. Strength and deformation characteristics of the ground.

A foundation should be designed so that the soil can safely sustain forces transmitted from the superstructure. The forces usually are horizontal base shear, overturning moment, and vertical force. Since the forces are usually applied for a short period and are dynamic in nature, ATC-3 suggests that the soil remain elastic under these forces.

Base shear is resisted by friction on the bottom surfaces of direct foundations and by lateral bearing on pile foundations. In deep pile foundations, passive pressure assists the lateral bearing capacity of the piles. A design check should also be made of the overturning moment applied to the foundations. Excessive settlement and separation of the piles should be avoided.

It is useful to restrain relative horizontal movement between footings by introducing connecting beams or slabs. ATC-3 suggests the use of $(A_v/4)N$ (Table 4-1) as the design horizontal force for connecting elements, with N as an axial force applied to the columns. In buildings with many

**TABLE 5-1 Standard Values
of Friction Coefficient**

Coarse soil without silt	0.55 ($\phi = 29°$)
Coarse soil with silt	0.45 ($\phi = 24°$)
Silt or clay	0.35 ($\phi = 19°$)

spans, relative vertical movement between footings may also be significant. In this situation, connecting beams or slabs should be strong enough to resist possible out-of-plane bending.

5.3.1 Direct Foundations

If there is a possibility of liquefaction, soil stabilization should be carried out. Otherwise, pile foundations should be used instead of direct foundations.

In estimating the horizontal shear resistance of foundations, AIJ (1974) recommends the use of 75 percent of the friction coefficient listed in Table 5-1. Values in the table have been determined by assuming shear capacities of tan ϕ and one-half of the compressive strength for sand and clay, respectively.

5.3.2 Pile Foundations

If there is a possibility of liquefaction, soil stabilization is useful. Otherwise, the vertical bearing capacity and horizontal resistance of piles should be estimated by assuming that the friction and horizontal soil reaction of those piles be zero in the range of potential liquefaction.

The soil's bearing capacity should be greater than the pile axial forces, which are caused by the vertical load and overturning moment of the superstructure. The base shear-resisting mechanism of piles is given as a combination of passive pressure, horizontal resistance of piles, and friction between the footings and the ground. If a pile is not driven deep, the effect of passive pressure should not be taken into account. Similarly, inclusion of the friction effect should be limited unless it is properly justified. For an estimation of pile horizontal resistance, the methods proposed by Chang (1937) and Broms (1964a, 1964b) are widely used.

In Chang's method, a pile subjected to horizontal force is treated as a beam on an elastic foundation. This method therefore requires values for the coefficient of horizontal reaction. Two approaches are available to find this coefficient: one by estimating it from the results of tests of long piles subjected to horizontal force, and the other by applying pressure in the borehole and estimating it from the relationship between the pressure and change of radius of the hole.

In Broms's method, piles are first classified into long and short piles, and for each case the stress distribution and ultimate horizontal resistance are computed by assuming a failure mode (Fig. 5-4). This method is simple and, in general, is in good agreement with experimental results (Broms, 1964a, 1964b, 1965). Ductility is required in the piles since they must sustain a combination of axial force, bending moment, and shear

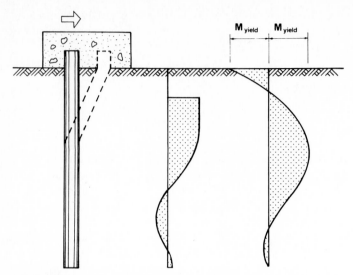

Figure 5-4 Deflection, soil reaction, and bending-moment distribution for a long restrained pile in cohesive soil by Broms. [*B. B. Broms, Lateral resistance of piles in cohesive soils,* J. Soil Mech. Found. Div., Am. Soc. Civ. Eng., *90(SM-2), 27−63 (1964).*]

during earthquake motion. ATC-3 presents recommendations for the design of vertical and stirrup reinforcement of piles. It should also be noted that connecting reinforcing bars are needed between the pile head and the cap.

REFERENCES

AIJ (Architectural Institute of Japan) (1974). *Design Specification of Building Foundation and Commentary,* AIJ, Tokyo (in Japanese).

ATC-3 (Applied Technology Council) (1978). *Tentative Provisions for the Development of Seismic Regulations for Buildings (ATC 3-06),* Nat. Bur. Stand. Spec. Publ. 510, Washington.

Broms, B. B. (1964*a*). Lateral resistance of piles in cohesive soils, *J. Soil Mech. Found. Div., Am. Soc. Civ. Eng.,* **90**(SM-2), 27−63.

——— (1964*b*). Lateral resistance of piles in cohesionless soils, *J. Soil Mech. Found. Div., Am. Soc. Civ. Eng.,* **90**(SM-3), 123−156.

——— (1965). Design of laterally loaded piles, *J. Soil Mech. Found. Div., Am. Soc. Civ. Eng.,* **91**(SM-3), 79−99.

Chang, Y. L. (1937). Discussion on "Lateral pile-loading tests" by Feagin, *Trans. Am. Soc. Civ. Eng.,* 272−278.

Ohsaki, Y. (1970). Effects of sand compaction on liquefaction during the Tokachi-Oki earthquake, *Soil Found.,* **10**(2), 112−128.

Seed, H. B. (1970). Soil problems and soil behavior, in R. L. Wiegel (ed.), *Earthquake Engineering,* Prentice-Hall, Inc., Englewood Cliffs, N.J., 227−251.

——— and I. M. Idriss (1971). Simplified procedure for evaluating soil liquefaction potential, *J. Soil Mech. Found. Div., Am. Soc. Civ. Eng.,* **97**(SM-9), 1249−1273.

Whitman, R. V., and F. E. Richart (1967). Design procedures for dynamically loaded foundations, *J. Soil Mech. Found. Div., Am. Soc. Civ. Eng.,* **93**(SM-6), 169−191.

6

SAFETY EVALUATION
AND STRENGTHENING
OF EXISTING BUILDING STRUCTURES

6.1 Evaluation of Seismic Safety

There is an increasing need to evaluate the seismic safety of building structures which were either designed with obsolete codes or damaged in previous earthquakes. Nevertheless, no systematic procedure has yet been established because of the short history of research into this subject.

Okada and others carried out one of the earliest evaluative studies (Okada and Bresler, 1976; Bresler, Okada, et al., 1977; Blejwas and Bresler, 1979). They developed a procedure to evaluate the seismic safety of reinforced-concrete low-rise buildings (up to five stories) and, in particular, school buildings. In order to judge systematically the seismic safety of many existing building structures in a short time, their method uses several levels of screening. First, the safety of a building is estimated by a simple analytic technique. If the building does not satisfy the specified safety requirement, it is again analyzed, this time by a more rigorous technique. This procedure, called *screening*, is repeated until the building clears the safety requirement. As the analysis becomes more rigorous, the magnitude of the safety margin set for the screening is relaxed. Safety is defined in terms of repairability following medium-size earthquake disturbances as well as noncollapse during destructive earthquake disturbances.

The basic procedure for the first screening consists of the following five steps:

1. *Structural modeling.* This step includes *(a)* checking the structural system (plan, sections, details); *(b)* checking load intensity; *(c)* checking material properties; and *(d)* examining the design method by means of drawings, design calculations, specifications, construction records, and field investigation.

2. *Analytical modeling.* It is assumed that the performance of the first floor controls the seismic safety of the building. First, each story is simplified to a single mass-spring system. Three types of failure mechanism are considered: bending, shear, and shear bending. Shear cracking strength, ultimate shear strength, and ultimate flexural strength are then determined according to the properties of the original building. A simplified method is used to calculate these strengths as well as the natural frequency. The response of the building is computed by a modal analysis including the first mode only. Instead of carrying out a time-history analysis, response spectra are used to evaluate performance with respect to strength and ductility.

3. *Safety evaluation: strength.* Linear response spectra are used for this purpose.

4. *Safety evaluation: ductility.* Nonlinear response spectra are used for this purpose. For safety evaluation, a table is provided (see Table 6-1).

5. *Combined evaluation of safety.* By combining the results of the evaluation of safety for strength and ductility, a final judgment is made as to whether or not the building is ultimately safe. If the judgment is

TABLE 6-1   Criteria Matrix for Judging Earthquake Safety of Reinforced-Concrete Buildings (First-Screening Stage)

Failure mechanism	0.3-g earthquake	0.45-g earthquake
Bending type (ductile)	Ductility factor μ* less than 2.0	Ductility factor μ less than 4.0
Shear type (brittle)	Shear cracking stage	Before shear failure stage†
Shear bending type	Shear cracking stage	Yielding stage‡

*Ductility factor = maximum displacement/yield displacement.
†Shear deformation at this stage is considered to be one-half of the ultimate deformation capacity ($\gamma_{ult} = 4 \times 10^{-3}$ rad).
‡Displacement at this stage is considered to correspond approximately to a ductility factor of 2.0 for the bending type.

positive, the building is taken to be safe; if the judgment is negative, it is taken to be unsafe. If the judgment is uncertain (in a broader range), further screenings are required.

In the second screening, the overall structural behavior of each story is estimated more precisely, and a time-history nonlinear response analysis is performed. In the third screening, the nonlinear analysis based on the nonlinearity of each member is adopted.

Okada later revised his evaluation procedure, and this revised concept is incorporated in the current Japanese guide for the evaluation of the seismic safety of building structures (JABDP, 1977a, 1977b; Aoyama, 1981). The Japanese guide presents three screening methods: the first screening, the second screening, and the third screening. A screening with a higher order requires a more detailed and sophisticated investigation to evaluate seismic safety. Analytic rigor is compensated for by the magnitude of the seismic index I_s, which is defined as

$$I_s = E_0 G S_D T \tag{6-1}$$

where E_0 = basic seismic index
G = geological index
S_d = structural design index
T = time index

Each of the three screening methods has particular procedures to compute these indices.

The site index G reflects the characteristics of the soil profile of the site. With very few available data, it is assumed to be unity. The structural design index S_d is a coefficient to represent (1) the irregularity of a building in plan and elevation, (2) story-stiffness distribution over the height of the building, and (3) lateral-stiffness distribution among the lateral-load-resisting components in each floor level. A value ranging from 1.2 to 0.43 is selected from a checklist matrix. In the time index T (from 1.0 to 1.5), long-term effects such as creep, shrinkage, settlement of the foundation, the effect of fire, etc., are taken into account. The basic seismic index E_0 is given as

$$E_0 = \phi CF \tag{6-2}$$

where ϕ = story index
C = strength index
F = ductility index

Equation (6-2) indicates that load-carrying capacity is a function of both strength and ductility. The ductility index F used in the first screening is listed in Table 6-2, whereas the indices for the second and third

**TABLE 6-2 Ductility Index
for First Screening**

Member	F
Column (height/depth > 2)	1.0
Short column (height/depth ≤ 2)	0.8
Wall	1.0

screenings, higher than those for the first screening, are classified in further detail. The strength index C is the lateral-load-carrying capacity of the floor level being considered, divided by the total mass above this level. A separate procedure is used to compute the strength index in each screening.

1. *First screening for C.* C is computed from the areas of walls and columns; i.e., only shear capacity is checked. Reinforcing bars in the columns and walls are ignored in the computation. Beams and floor slabs are considered to have adequate rigidity and strength.

2. *Second screening for C.* Both flexural and shear strengths are computed for columns and walls. As with the first screening, beams and floor slabs are considered to be rigid and strong.

3. *Third screening for C.* Compute column and beam ultimate moments, and take the smaller of the two as the maximum moment. These nodal moments then are distributed to the beam or column ends in accordance with their stiffness.

The story index takes account of the story level being considered and is given as

$$\phi = \frac{n + 1}{n + i} \tag{6-3}$$

where n = the number of stories in the building and i = the number of the story level being considered.

Chapter 13 of ATC-3 stipulates two steps for the evaluation of seismic hazard in existing buildings (ATC-3, 1978):

1. Qualitative evaluation

2. Analytical evaluation

Qualitative evaluation, which involves examination of design documents and field inspection, results in one of the following three decisions:

1. The building conforms to the provisions.

2. The building does not conform to the provisions.

3. Conformity cannot be determined by qualitative evaluation, and analytical evaluation is required.

To arrive at the decision, various conditions are considered. The capacity ratio r_c is introduced:

$$r_c = \frac{V_{as}}{V_{rs}} \tag{6-4}$$

where V_{as} = seismic shear-force capacity computed for the existing building

V_{rs} = seismic shear force that the existing system would be required to resist to meet the requirements of these provisions for a new building

Minimum acceptable values of the capacity ratio are specified. If the capacity ratio of a building is less than as specified, the building must be strengthened or demolished immediately. If the capacity ratio is greater than unity, the building is evaluated as safe according to current codes. If the capacity ratio is between the required minimum and unity, the building needs to be either demolished or strengthened within a specified period.

6.2   Repair and Strengthening of Existing Buildings

A building damaged by earthquake must be repaired in a manner which ensures that its original strength level is achieved or exceeded so that it can survive in future earthquakes. Chapter 14 of ATC-3 (1978) deals with repair techniques for various structural and nonstructural elements, as well as foundations, for construction in steel, reinforced concrete, wood, prestressed concrete, and masonry. Repaired buildings should satisfy the code requirements. The repair, however, may be very costly. Final decisions on whether or not to repair and, if so, how to repair, need to be made in the light of overall economy.

Shear walls, moment-resisting frames, horizontal diaphragms, and connections are most susceptible to damage, revealing common failure patterns. Typical methods of repair and strengthening are:

1. To remove damaged elements and replace them by new elements

2. To thicken, enlarge, or strengthen elements

3. To add new shear walls, vertical bracings, and columns to the structure

4. To convert shear connections to moment-resisting connections

5. To reduce the mass of the structure by removing upper stories

6. To examine the dynamic characteristics of the repaired (strengthened) structure

The effectiveness of such methods of repair and strengthening can be greatly increased by the use of the original design documents and records of construction.

1. *Structural-steel components.* In order to check the material strength or welding grade of steel in a building structure, we must call on codes and standards that were used in the design as well as the construction of the structure. It may therefore be necessary to go back to obsolete codes which were current when the building was constructed. Further, the fatigue history and reduction in cross-sectional area caused by corrosion must be measured. Nondestructive testing, including ultrasonic, radiographic, magnetic-particle, and fluorescent magnetic-particle testing, is also effective in inspecting steel assemblages and members. In addition to such nondestructive testing, it is often necessary to perform tests of material samples cut from the members.

Retrofit and strengthening of a steel structure can be achieved by various methods which avoid changing the structural system:

a. Replacing existing bolts and rivets by high-tension bolts

b. Welding previously unwelded connections

c. Reducing unsupported length

d. Increasing cross-sectional areas

e. Replacing members by others with higher strength

In some cases, it is better to change the structural system, e.g., from timber to reinforced-concrete or steel floor slabs or from steel braces to shear walls. Whenever new components are added to a structure, they should be arranged in a symmetrical manner.

2. *Reinforced-concrete components.* Design documents and construction records are most useful for checking the material properties of reinforced-concrete structural members. The location of reinforcing steel bars should be determined, if necessary by measurement. As repair material, *(a)* shotcrete (also called gunite) is often used, *(b)* epoxy resin is used to repair concrete cracks and cavities, and *(c)* epoxy mortar can fill large cavities more effectively than epoxy resin. Besides these materials, *(d)* gypsum-cement concrete, *(e)* portland-cement concrete, *(f)* quick-setting cement mortar, and *(g)* preplaced aggregate concrete are used for repair work.

Various methods of repair are employed:

a. If an opening is 6 mm or less, pressure-injected epoxy is convenient.

b. If cracks are large or the concrete is completely shattered, the epoxy-injection technique is no longer suitable, and shotcrete is more appropriate. Additional reinforcing bars can be introduced if adequate anchorage can be provided. Damaged reinforcing bars can be repaired by butt welding, lap welding, or splicing. If strengthening is needed, care must be taken that adequate force distribution is ensured in the strengthened system.

Two typical procedures for strengthening are:

a. Increase cross-sectional areas with additional main and confining steel bars.

b. Thicken shear walls by adding a layer of reinforced concrete to their surfaces.

Many studies have been made of the connections of concrete placed to strengthen existing structures (Sugano, 1981; JABDP, 1977*b*; Kahn and Hanson, 1977; Higashi, Ohkubo, and Fujimata, 1977; Yokoyama and Imai, 1980; Guangqian and Lian, 1981).

3. *Precast-concrete and prestressed-concrete structures.* In precast- and prestressed-concrete structures, methods of repair and strengthening are essentially the same as in reinforced-concrete structures. Material properties and the intensity of prestressing in tendons and strands must be measured carefully.

4. *Wood.* Methods of analysis and design for the repair or strengthening of existing wood members and frames are identical to those for new construction. Initially the quality of the materials is investigated, and some parts may be replaced if necessary. In some cases, it may be advisable to increase the resistance of a building to seismic forces by design modifications.

5. *Masonry.* In masonry structures, both frames and masonry systems must be inspected for cracks and material properties. Electromagnetic testing equipment can measure the location of steel bars, while sonic testing is useful to check crack development. Repair by shotcrete, by dry packing with portland-cement aggregate mix, or by the injection of mortar and epoxies is effective. Strengthening by prestressing, by bonding reinforcing mesh to masonry surfaces with plaster, by bonding glass-fiber reinforced-cement plaster to masonry surfaces, or by attaching braces and stiffening elements of metal or other material to masonry components by means of bolts or anchorages is effective for masonry systems.

In Japan, a guide for the repair and strengthening of existing buildings has been enforced since 1977 (JABDP, 1977*b*). The content of this guide is summarized below.

Strengthening can be achieved by increasing bearing capacity, increasing ductility, or increasing both strength and ductility.

1. *Increasing strength.* This method is most often used when a building structure has insufficient strength and it is not possible to upgrade its ductility.

a. Wing wall attached to columns and shear walls. (See Fig. 6-1.) Wing walls can be constructed in situ, although precast wall panels can also be used. In either case, the connection with existing columns or walls must be effective and is achieved by means of shear cotters, anchor plates, and chemical bonds.

b. Steel bracing. Steel bracing is an advantageous strengthening system since it adds only minimal weight to the original structure. Significant openings are maintained even after the braces have been introduced. As in sidewall strengthening, care must be taken in connecting the braces to the frames.

c. Buttresses. Buttresses are suitable for strengthening if space is available outside the building.

2. *Increasing ductility.* If the strength of a building structure is insufficient and if strengthening, e.g., with bracing and sidewalls, is not feasible, ductility improvement is a possible alternative. For example, brittle shear failure of columns can be altered to ductile flexural failure by arranging shear reinforcement in the columns. Several typical procedures are:

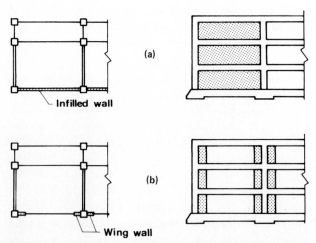

Figure 6-1   Strengthening by placing walls. *(a)* By infilled walls. *(b)* By wing walls. [*From S. Sugano, Seismic strengthening of existing reinforced concrete buildings in Japan,* Bull. N.Z. Nat. Soc. Earthquake Eng., *14(4), 209–222.*]

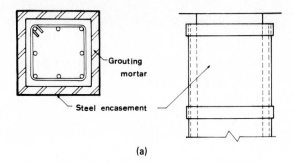

(a)

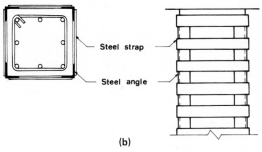

(b)

Figure 6-2   Column-strengthening detail. *(a)* By steel en-
casement. *(b)* By steel straps and angles. [*From S. Sugano,
Seismic strengthening of existing reinforced concrete buildings in
Japan*, Bull. N.Z. Nat. Soc. Earthquake Eng., *14(4)*,
209–222.]

a. Surround a brittle column by steel plates with some space set
between the column surface and these plates, and then fill the space with
mortar (Fig. 6-2*a*).

b. Place steel angles at the corners of a column, and connect these
angles by tie plates (Fig. 6-2*b*).

c. Surround a column by a welded-metal sheet, and place concrete or
mortar on the sheet.

3. *Increasing strength and ductility.* A building structure may have a
discontinuity in story stiffness. As a typical example, the first story may
have a piloti, while the upper stories have stiff shear walls. In this case,
continuity of stiffness can be achieved by stiffening the first story with
shear walls.

REFERENCES

Aoyama, H. (1981). A method for the evaluation of the seismic capacity of existing reinforced concrete building in Japan, *Bull. N.Z. Nat. Soc. Earthquake Eng.*, **14**(3), 105–130.

ATC-3 (Applied Technology Council) (1978). *Tentative Provisions for the Development of Seismic Regulations for Buildings (ATC 3-06)*, Nat. Bur. Stand. Spec. Publ. 510, Washington.

Blejwas, T., and B. Bresler (1979). *Damageability in Existing Buildings*, Rep. EERC 78-12, Earthquake Eng. Res. Center, College of Engineering, University of California, Berkeley.

Bresler, B., T. Okada, D. Zisling, and V. V. Bertero (1977). *Developing Methodologies for Evaluating the Earthquake Safety of Existing Buildings*, Rep. EERC-77-06, Earthquake Eng. Res. Center, College of Engineering, University of California, Berkeley.

Guangqian, H., and W. Lian (1981). On aseismic strengthening of existing reinforced concrete frames. *Proc. U.S./PRC Workshop Seismic Anal. Des. Reinf. Concr. Struct.*, Ann Arbor, Mich., 215–226.

Higashi, Y., T. Endo, M. Ohkubo, and Y. Shimizu (1980). Experimental study on strengthening reinforced concrete structure by adding shear wall, *Proc. Seventh World Conf. Earthquake Eng.*, Istanbul, **7**, 173–180.

———, M. Ohkubo, and K. Fujimata (1977). Behavior of reinforced concrete columns and frames strengthened by adding precast concrete walls, *Proc. Sixth World Conf. Earthquake Eng.*, New Delhi, **3**, 2505–2510.

JABDP (1977a). *Standard for Seismic Capacity Evaluation of Existing Reinforced Concrete Buildings*, Japan Association for Building Disaster Prevention, Tokyo (in Japanese).

——— (1977b). *Design Guidelines for Aseismic Retrofitting of Existing Reinforced Concrete Buildings*, Japan Association for Building Disaster Prevention, Tokyo (in Japanese).

Kahn, L. F., and R. D. Hanson (1977). Reinforced concrete shear walls for aseismic strengthening. *Proc. Sixth World Conf. Earthquake Eng.*, New Delhi, **3**, 2499–2504.

Okada, T., and B. Bresler (1976). *Strength and Ductility Evaluation of Existing Low-Rise Reinforced Concrete Buildings—Screening Method*, Rep. EERC-76-1, Earthquake Eng. Res. Center, College of Engineering, University of California, Berkeley.

Sugano, S. (1981). Seismic strengthening of existing reinforced concrete buildings in Japan, *Bull. N.Z. Nat. Soc. Earthquake Eng.*, **14**(4); 209–222.

Yokoyama, M., and H. Imai (1980). Earthquake damage at Izumi High School in 1978 Miyagiken-Oki earthquake and methods of repair and strengthening, *Proc. Seventh World Conf. Earthquake Eng.*, Istanbul, **4**, 81–88.

CONVERSION OF MEASUREMENTS

$1\ \mu m = 39.37\ \mu in$ $1\ km = 0.62\ mi$

$1\ mm = 0.03937\ in$ $1\ m^2 = 1.196\ yd^2$

$1\ cm = 0.3937\ in$ $1\ MPa = 145.04\ lbf/in^2$

$1\ m = 3.2808\ ft$ $1\ gal = 1\ cm/s^2$, or $0.337\ in/s^2$

NAME INDEX

SUBJECT INDEX

About the Author

A graduate of Tokyo University (D. Eng. in the Department of Architecture, Faculty of Engineering), Professor Minoru Wakabayashi is the author of *Design of Steel Structures* and *Earthquake Resistant Structures* and of more than 250 technical papers on steel, reinforced-concrete, composite steel and concrete, and masonry structures. He has taught in the Department of Architecture and the Graduate School of Engineering, Kyoto University, and has served as director of the Disaster Prevention Research Institute of Kyoto University. In addition, he has examined the safety of many proposed building structures for various governmental commissions.